Product Management

Product Management

The Art and Science of Managing Network and
Communications Industry Products

Ishrat Nadeem Zahid

To order additional copies of this book, contact:
Xlibris Corporation
1-888-795-4274
www.Xlibris.com
Orders@Xlibris.com
134638

This book, my first, is dedicated to my mother who took care of me day and night so that I could become what I am and to my father who worked hard so that I could be well educated.

About the Author

Ishrat Nadeem Zahid, known as simply Nadeem to his friends and colleagues, is a communications industry veteran. Over his eighteen years career, Nadeem has held engineering, product management/ marketing, and strategy-related positions with leading technology companies, mostly in the Silicon Valley, where he has built, launched, and marketed several successful hi-tech products and solutions exceeding estimated $2.5 billion in net revenues.

Nadeem started his journey toward the hi-technology career as a software engineer with Alcatel Telecom., working on traditional voice telephony systems. He then worked for Lucent Technologies on shaping up the wireless and cellular technologies for voice and data communications. It was really at Cisco Systems where he took off on the emerging wave of disruptive technologies at the time such as Voice over IP and modern data networking. At Cisco, Nadeem worked on several products and technologies in the enterprise, data center, and service provider areas and also transitioned to a product manager role. After several years at Cisco, Nadeem joined Foundry Networks and expanded his responsibility as a product line manager to manage broader enterprise and data center product portfolio. After Brocade Communications acquired the Foundry, Nadeem managed the product marketing and go-to-market strategy for all IP/Ethernet products portfolio. At Juniper Networks, as director of product management, Nadeem managed major strategic alliances and OEM/resell business with partners like IBM, Dell and Ericsson. Lately at Extreme Networks as director of product-line management and marketing, Nadeem introduced flagship enterprise and cloud data center products with disruptive technologies.

Nadeem holds bachelor's of engineering (BE) degree in electronics engineering from the NED University of Engineering & Technology, Pakistan. He also earned his master's of science (MS) degree in telecommunications from Boston University, USA. In addition, he earned several professional certifications including a project and risk management certificate from Massachusetts Institute of Technology (MIT), USA, as well as Cisco Certified Networking Associate (CCNA), Cisco Certified Network Professional (CCNP), and Cisco Certified Internetwork Expert (CCIE) certificates.

In addition to a busy career, Nadeem has had a very active lifestyle. He earned his private pilot license and flew airplanes. He has enjoyed composing music, painting, golfing and photography. These days, in his spare time, Nadeem enjoys time with his family. At the time of writing this book, Nadeem lives in San Ramon, California, with his wife, son, and a daughter.

Contents

Preface

From time to time, I get approached by ex-colleagues and other professionals who seek advice on if and how they can transition into a product management role from other roles such as engineering. The reason they approach me is because they have witnessed me making such a transition successfully several years back, and then they have also witnessed me building and delivering some of the most successful products in the network and communications industry. Usually, they have some common reasons behind this motivation, but not always the right reasons. Some of them are bored with what they do, some do not see much career growth, and some starve to learn the business side of the technology, some want to satisfy their entrepreneurial spirit, and yet some want to do it just because they see others do it.

Regardless of what the source of motivation for such a transition is, most of those people are unfamiliar with what product management is all about, what it takes to become a successful product manager, and what it takes to sustain it. They are also unfamiliar with how this transition could affect their personal lifestyle. I would usually meet with the person wanting to take a "brain dump" from me over lunch or coffee, and our conversation would begin with me asking this question "Why do you think this is the right thing for you?" If the person really knows the answer, then we can have more conversation. Otherwise, it is a fairly short meeting. In other cases, I am approached by the product managers who are on a learning path. They have lots of unanswered questions that no business school has taught them. The answers can come only from the real world experience.

I have found that there is really a shortage of good knowledge transfer about the product management profession, especially in the hi-technology industry, and what it is all about. That is where the idea of writing this book came from. This book is the essence of my several years of experience in product management and marketing. It answers the questions for someone who is considering the product management career, and it serves as a crash course for existing product managers. I hope, even the seasoned product managers could benefit from the best practices outlined in the book. Since most of my experience has been in the field of networking, I have used it for examples throughout the book while trying to stay at generic level overall so that the book can also be useful for other than network products.

With an extremely demanding career and full family life, this was a huge undertaking. I want to thank my wife and my little ones who let me "steal" the time to get it done. However, I am happy of the outcome. Thinking that this effort of mine could help many, who need the required knowledge transfer and mentorship, is very satisfying.

Ishrat Nadeem Zahid

June 2013
San Ramon, California, USA

Great products are the most important thing for a hi-technology company—period.

CHAPTER-1

PRODUCT MANAGEMENT OVERVIEW

What is Product Management?

It may not be surprising for us to know that most of the things around us require some sort of management. For example, our houses require ongoing management that includes maintaining the house in a healthy state, paying the mortgage, taxes, and bills on time, and even putting some improvements or features in the house to increase its net worth. Our cars require management too, including oil changes, routine maintenance, car washes, registration renewals, and insurance payments. Our kids require management too. One has to make sure they are fed well, their health is monitored and maintained in the best condition, that they are developing the right behaviors, they are well educated, and their activities, social life, and other aspects are taken care of. Generally, one will only abandon managing something if one stop caring about it or it does not mean anything.

If we were to view and treat the subjects above as some sort of products, then we can say that almost all products require management, hence the term *product management*. Although the examples of the products quoted are mostly not for business reasons or necessarily to make a profit, in reality most products are built to solve certain problems or to perform certain jobs and selling them for doing so. Therefore in most cases, product management is tied with managing products to generate business out of them. In professional context, a *product* is a physical or virtual entity produced by labor and designed to solve a certain problem, to function in a certain way, and to produce a certain output for monetary purposes and as demanded by the product users or customers. Therefore, the product has to be usable and valuable for the end users in order for them to pay for it. A *customer* is someone paying to use the product or buying it on behalf of a user.

> *A product is a tangible or intangible entity produced by labor and designed to solve a certain problem, to function in a certain way, and to produce a certain output for monetary purposes and as demanded by the product users.*

A product is usually designed to solve a particular problem or serve a particular purpose; it is usually designed for particular needs of a particular set of users who will likely use it. It is therefore designed per the requirements those users have and per the specifications and attributes that those users like the product to have. A product can be a tangible item that you can see, touch, feel, and use in a certain way. But it can be more than that. With the growth of Internet, a product now can also be something you cannot touch, but you can still see it and use it in a defined way. Many software products (applications) that are sold and used in the cyber space are examples of this. A product could even be a service. Although that is not entirely true as we will explore later in the book, but at this point of introduction, a service could be considered and treated as a product. A service is an intangible entity. It cannot be seen or touched but can only be experienced. A product therefore is not just a random idea and an accident but rather well thought out and designed to serve a particular purpose and need. Depending on how successful the product is in meeting those needs, it will determine the success or failure of the product and how much users are willing to pay for it.

The *product management* in this context involves managing the full life of such a product, including starting with a concept or idea, identifying market opportunities for it, defining its intended functionality and usage, defining its internal details, getting it built, positioning it in contrast to a any competitive products, setting its selling price, pushing it in the market, generating revenue out of it, adding ongoing improvements to it, and eventually pulling it off the market and terminating it. This cycle is repeated over and over with products during their life from the cradle to grave and is referred to as the *product life cycle management*. Most of these activities can be summarized into different stages of the product life cycle, namely planning, execution, launch, sustaining, and termination.

> *The product life cycle consists of a product life as it moves from cradle to grave including product planning, execution, launch, sustaining, and termination phases.*

We will later on examine different phases of this life cycle in greater details. Therefore in summary, *product management* means defining, building, and growing products in relation to the end user requirements and for monetary purposes.

> *The product management means defining, building, and growing products in relation to the end user requirements and for monetary purposes.*

There are three important points to emphasize here. The first point is that the products always need to be designed and built with end user and end use in mind. This can be the key difference alone between the success and the failure of a product. The *end user* and the *customer* (who purchases the product) are usually the same or, for the sake of discussion, it is safe to assume so, and we can use the term interchangeably; however, that is not always the case. We will further discuss this distinction later under channel sales. If a product is not designed in a way the customer will like to use it, or it is not easy enough to use, or it does not provide the sort of functionality or performance the customer expects out of it, then it will be rejected by the customer because of poor *customer experience*. We will explore the customer experience later on in more detail as it is getting more important with time.

Second, the emphasis on building products for monetary purposes that can generate revenue and profits for the company is important. Building products that do not sell does not make much sense. This is an important point to focus on. Many product managers in hi-technology industry evolve from engineering background and perform an excellent job in terms of designing and building great products functionality wise. However, they sometimes fail to consider the business and usability side of the technology, and these great products end up being a failure in the marketplace. Instead of making any money, they cost lots of *research and development* (R&D) dollars to the company. Therefore, the *business thinking* is as important as the *technical thinking* in the product management profession.

Third, it is important to note that no product lives forever. As the market trends and user expectations change underneath, the older generation products have to be phased out, and newer generation of products need to be introduced. As the technology evolves in terms of components used in the products as well as the manufacturing processes and user expectations evolve, the newer generation products are usually much better in term of functionality, efficiency, cost, and other aspects compared to their predecessors. However, since older generation products are usually cleaned up of defects over period of time and are much more stable and predictable in terms of behavior, balancing the quality with innovation is a delicate judgment call, and we will examine this very important aspect as well in later chapters in more detail. Any negative changes when migrating from older to newer products resulting in poor customer experience are usually not good as customers like going only upward in experience.

Role of a Product Manager

As explained earlier, there has to be someone who can take care of all aspects of a product throughout its life and take on the responsibilities involved in managing the product. That person is called a *product manager*. Simply put, a product manager manages and supervises a product. The product manager handles the strategic and tactical duties related to different phases of the product life cycle. In a way, the product manager is the CEO of the product and determines the fate of the product by making decisions about it on daily and weekly basis, throughout its life. The product manager is the link between the business and the customers, and without this link, it would be extremely hard to generate much of the business from the customers.

> *A product manager manages a product throughout its life cycle.*

Being a product manager is a tough job, yet it can be very exciting and rewarding. The product manager deals both with inbound and outbound sides of the company. On the inbound side, the product manager works with almost every functional group in the company to manage a product. This includes working cross-functionally with engineering, manufacturing, compliance, operations, documentation, marketing, sales, legal, and other teams as well the top-level executives. Most of the time, the product manager has no authority over those functional teams and has to get the work done through them, which is an ongoing challenge and a skill in itself. Working in such a structure is usually referred to as a *matrix organization.*

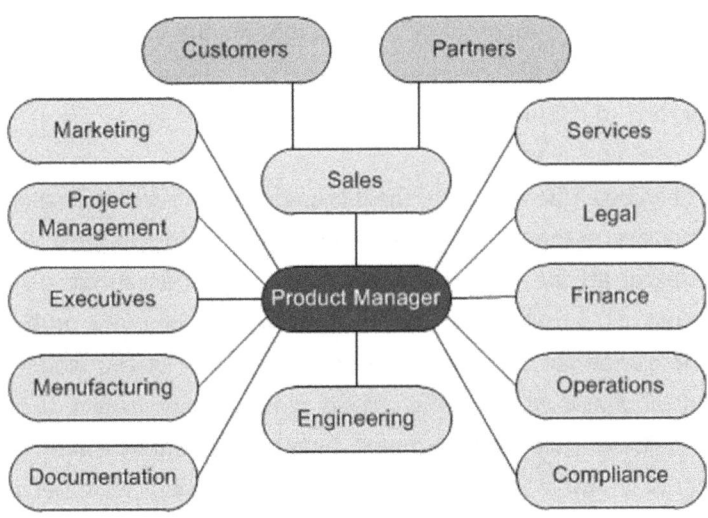

Diagram 1: A Product Manager's Interaction Scope

Working in matrix organizations is not easy and requires strong teamwork, collaboration, and influencing skills. It is absolutely critical that the product manager is empowered by the top management to make everyday decisions. The product manager exercises influence as the fundamental tool to get things done. On the outbound side, the product manager works closely with the customers, usually through the sales team, to understand their problems and pain points. The product manager then bridges the two worlds together by translating the customer needs in the form of *requirements* and acting as the customer face and voice for the inside world. Writing clear requirements is critical to minimize the gap between what customers expect and what is actually delivered at the end, and it is an art as much as it is a science. We will examine this topic in great details later due to its importance.

One could also raise the questions of why a product manager is needed to communicate with the customers for requirements gathering, and why engineering or other functional teams cannot directly interact with the customers and do the same. There are several reasons for not doing so. First of all, there are usually delicate issues and protocols for communicating and interacting with the customers such as what to expose and what not to expose to customers as well as relationship management that inbound teams are not trained or focused on. Moreover, there can be legal issues as customers can hold the company accountable for the statements made that could be interpreted as commitments. For this reason, companies do not allow everyone to communicate with the customers.

Second, it can be a distraction for the inbound teams to add this responsibility to their charters on top of their core responsibilities. Most of the internal team members excel at one thing such as engineering and cannot evaluate the requirements in a business context. In other words, they lack both technical and business experience at the same time. They are not best suited to filter the useful requirements from the useless requirements. In fact, it can cause great confusion and chaos as different functional groups debate over which requirements are more important than others, and it can soon become even more challenging as people debate which customer's requirements are more important. Therefore, someone needs to filter, consolidate, and prioritize the requirements from multiple sources for the inside teams.

Other than the requirements gathering and communication, the product manager is involved supervising the new product development process and correcting any deviations or expectations and providing clarity. There are many other activities that the product manager carries out after the requirement gathering is complete that cannot be performed by engineering, operations, or other functional groups. Such activities include managing the full life cycle of the product involving tasks such as market analysis, business case, forecasting, pricing, positioning, product roadmap, competitive strategies, and sales enablement. During this time, the product manager decisions are considered authority over anyone else, after executives. This also means that the product manager is accountable for any mistakes made and for any business loss resulting from those decisions. It is therefore a highly visible but responsible position. Generally, the product managers have the *profit and loss* (P&L) responsibility for the product and as measurement and effectiveness of their success and contribution toward the company business. This is one of the most important positions in a company. The remaining sections of this book go in details into all aspects of product management and the responsibilities of a product manager.

Product Line Management

Now that we understand a bit what product management is and what the role of a product manager is, it is useful to explain the difference of a product manager and a product line manager. Whereas the product manager manages a single product, a *product line manager* (PLM) has the responsibility to manage multiple products, usually a complete product family or multiple product families. The product line manager can either directly manage multiple products by itself or by managing a team of product managers who then manage their products.

> *A product line manager manages more than one product or product families.*

The product line manager not only keeps an external perspective in terms of what needs to be built to compete with the competitive offerings but also need to maintain an internal perspective for intra-product positioning and competition. The product line manager deals with taking the full product line or portfolio forward as a complementing solution against the competition and not let the internal products cannibalize each other. Maintaining clear positioning among different products and have clear go-to-market strategy plays a key role in maintaining successful product lines.

Product Management versus Product Marketing

Sometimes, the product management and product marketing terms are used interchangeably. Technically, they are two different roles as we will explore later in the book. The product management role, as we have discussed earlier, is more of an inbound role working with multiple functional groups to build, deliver, and manage a product. It is both a strategic and tactical role and is centered on the technical skills. The product marketing is more of an outbound role, responsible for promoting and evangelizing a product or product family. This includes activities like managing a product launch, creating product messaging, originating product collateral, as well as planning and participating in public events to promote the products. The product marketing works with external stakeholders such as sales, business development, industry analysts, and media. Since the product marketing is closely related to the product management, sometimes it is also under the charter of a product manager, and the lines are blurred. In larger companies and especially companies who take product marketing seriously, it is a focused and separate role under the domain of a *product marketing manager* (PMM).

> *Product marketing is focused on marketing the products versus building them.*

Product Life Cycle

As introduced above, every product, small or large, simple or complex, has a limited life. It starts with an idea and finishes with an end of sale and support. There are some key stages or phases, through which almost all products go through. Each and every phase in the life of a product is directly supervised by the product manager. At a high level, we can define those stages or phases as *planning, execution, launch, sustaining,* and *termination* phases. The next chapters will explore those key phases in detail and explain important considerations as they relate to those phases. This chapter introduces those phases at high level.

Diagram 2: Product Lifecycle Phases

The planning phase, as the name says, deals with putting together a detailed plan on what, why, how, and when. What is the market need? What is the solution? What will be the product? Why it should be developed? How it should be developed? How much it would cost? How it will be positioned and sold? How much money would it make? And, When it should be delivered? Those are some of the questions that must be answered in the planning phase as clearly as possible. The planning phase consists of several activities including a product idea or concept, requirements definition, and high-level product design. We will explain and explore those tasks in detail in next chapters.

The execution phase deals with the actual development of the new product as planned. Given that the planning was done carefully and thoughtfully, execution should not hit any major obstacles or surprises. But that is usually not the case. Therefore, the execution phase is also the time to remove any unforeseen obstacles as they arise. *Time to market* (TTM) of a product heavily relies on managing the execution phase. A well-executed execution phase results in timely delivery of the product, and a not-well-executed execution phase results in a delayed product delivery that can severely cripple the success of the product.

The launch phase deals with introducing the product in the market place once it is ready. No matter how great the product is, it must be marketed and advertised so that the target customers can find out about its availability and can plan to buy the product. Launch phase deals with tasks such as publishing the product collateral on company web pages, showcasing the new product in trade shows and public events, and flooding the marketing messages to potential customers so that they can be motivated to buy the product. More launch-related activities are discussed in the chapters later.

After a product has been launched, it needs to be sustained. Once a product is sold and deployed, customers usually come back and ask for more features and improvements in the product, as they get more familiarized with the product and how they can benefit more from it. Moreover, the product use in real *production environment* will uncover the defects that are usually very hard to uncover during internal *quality assurance* activities. Therefore, those defects will need to be fixed and released to the paying customers in the form of new *revisions* of the product or in the form of *software updates*. Moreover, supply chain and orders will need to be managed and fulfilled in a timely manner. It will also need to be continued to be marketed, and ongoing business opportunities will need to be developed so that the product continues to sell in anticipated volumes. The sustaining phase deals with the above-mentioned tasks.

Finally and eventually, there will be a termination of every product after it has lived its life and is either getting obsolete or being replaced by another, usually better, product. Terminating the life of a product, or *end of life (EoL)* as it is called, also requires careful planning and messaging. How the older product could be cannibalized by the newer product, and how to shift the revenues over. How much it costs to notify the customers in advance who are using the older product. How to help the existing customers migrate on to the newer product to transition business and long-term revenue prospects without causing unhappy customers and lost business or reputation. These are some of the issues associated with the end of life of a product. We will explore all these topics in detail in the later chapters.

Product Management as a Career

Product management is not a job; it is a career. Product management can be a great and exciting career, but the question is that what is the basic qualification to become a product manager? Many product managers come from engineering background in the hi-technology industry, and there are few well-known reasons for that. The most common one is that after spending years in engineering roles, engineers get bored. They feel that there is nothing more they can do outside of their defined roles, and they look for more versatile career tracks that have good mix of technical and business tasks to work on, in order to develop new skills and advance their careers. Some may feel stuck in their careers and do not see a line of sight to grow and excel in their existing careers and find the need to switch tracks. They may feel myopic in terms of business and market awareness of the very products they build and may want to develop broader business and entrepreneurial skills.

Most of these technical experts know their functional areas deep down but lack a broader and big picture view of things, such as where and how the products they work on, are used? What are major industry trends, and how they shape up and influence the work they do? What the end users think about the products they develop? Yet another reason is increased outsourcing of engineering work to overseas, and many engineers find it almost critical to switch careers for long-term job security reasons. These are some of the reasons that motivate engineers to step out of their conventional roles and enter new roles while carrying forward their technical foundation and putting it to use in new ways.

Engineers are not the only ones switching careers to product management. Many product managers come from other backgrounds such as *technical marketing engineering* (TME), *product engineering* (PE), or *sales engineering* (SE) roles. Yet some start their career straight out of a business school after their MBA (master's in business administration) degree. No matter what the reason could be or what the background is, the more important question is that should one make a transition to product management in the middle of one's career? Or before we discuss that, should one even start in this career after school? Before answering those important questions, there are few other questions that must be answered first. The most important question for someone to ask himself or herself is that where does he or she want to head in the long run? The product management career usually leads to a *general management* (GM) position in the long run. General management is a business operations role that looks over many functional areas, and usually, a general manager manages both revenue and costs of the particular business entity within a company with full P&L responsibility.

Therefore, someone who likes to ultimately step into an executive position to make a difference in the top and the bottom line of the company's income statement by managing both the technology and business teams to deliver products, services, and solutions, should consider starting in the product management role. In very large companies, there are usually multiple general managers for multiple business entities usually called *business units* (BUs). In which case, there can also be a *chief development officer* (CDO) role to look over multiple business units and general managers. Alternatively, if someone wants to simply learn with an entrepreneurial spirit how to design, develop, and deliver products and how to generate business out of them, to eventually start their own business, should also consider spending few years in product management role.

> *A general manager (GM) looks over many functional areas and carries the profit and loss (P&L) responsibility by managing revenues and costs of the business.*

It should be well understood as explained earlier that the product management is a very high-pressure job. One will be busy every day, and there is no time to get bored. Usually, it is a team of one. Every day in the life of a product manager is a different day and brings different challenges to solve and different set of decisions to make. A product manager is the most responsible and single go-to person for all things products, and therefore, all worries belong to him or her. He or she is the single caretaker of the product. It is not a role for the fainthearted. As we will discuss in detail in the following chapters, the product managers face higher-level scrutiny for the kind of decisions they make which impact the success and failure of the entire product and sometimes the entire business. However, this degree of challenges also brings lots of excitement to the role. This is an excellent position to develop good understanding of multiple functional areas of the company and how they interact and contribute toward the common goals and how the business is run. It also helps build new relationships within and outside of the company. The role is highly visible to the top company executives. Good product managers are well respected and rewarded, and no one wants to lose them.

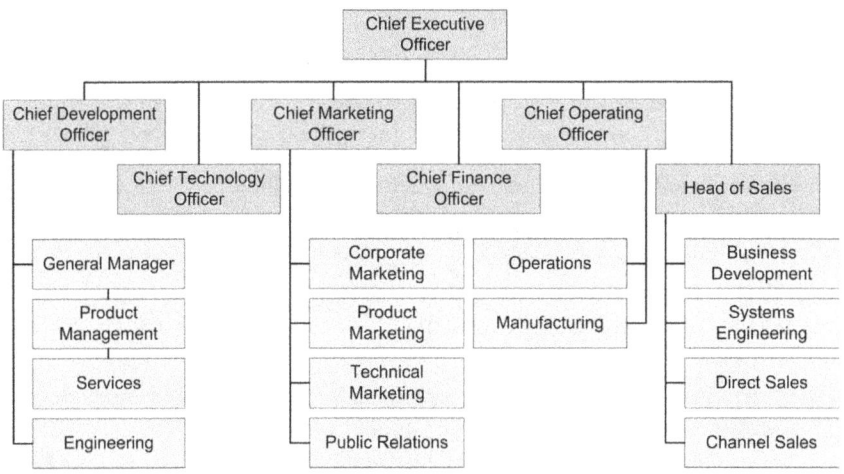

Diagram 3: Typical Org Chart of a Hi-Tech Company

How to Become a Product Manager?

Given that one has made up the mind to move forward with product management as a career as a mid-career transition, the next challenge is how to make it happen. Simply wanting to become a product manager does not help. Due to the high visibility, responsibility, and accountability of the role, the barrier to entry is usually high, and there is no lack of candidates. Moreover, there is only handful of product management positions in a company compared to other functional area such as engineering or operations. The hiring manager, typically a director or VP/GM of product management, looks for a certain type of personality in addition to the required credentials and experience. Below is a typical job requirement description advertised for a product manager position in a networking company.

EXAMPLE 1

Product Manager Opportunity with Leading Networking Company

- Ability to synthesize complex technical and market issues, develop solutions for those challenges, and gain acceptance and commitment from others.
- Manage projects from product definition through launch.
- Define and drive key product features, identify core product target market, and create product messaging and positioning.
- Develop and give EBC presentations to customers and taking customer feedback.
- Understand and communicate competitor's positioning and strategies and develop competitive strategy and positioning.
- Work with finance, operations, and sales on detail forecasting and predictive analysis.

Technical background required includes

- An in-depth understanding of networking products and services.
- Extensive technical, industry, and business knowledge of the networking space.
- Experience working with networking technology in production environments.
- Financial modeling.

In addition, the following skills and experience are highly desirable:

- Excellent written, verbal, and presentation skills. Strong listening, negotiation, project management, and decision-making skills.
- Superior interpersonal skills. Ability to build relationships at all levels with sales/marketing staff, channels, resellers, and customers.
- Understanding of and experience in applying complete life cycle marketing, including definition, launch, transition, or end of life, through both direct sales and indirect channels.
- Work independently as well as in team environments to complete projects using judgment and experience.
- Flexibility in a fast paced, ever changing, deadline-driven environment.

Experience Required:

- Typically requires BSEE/CS or finance, marketing or equivalent, plus 10+ years related experience in marketing networking technologies and applications.
- MBA or advanced degree strongly preferred.
- Extensive presentations skills.

Many engineers think that going back to school for a full- or part-time MBA will get them into product management career. That is not entirely true. Although having an MBA degree is desirable for the product management role, it is not good enough by itself or may not be a must-have. What the hiring managers look for is more of an aptitude, personality type, relevant experience, broad knowledge, business acumen, and soft skills, and on top of that having an MBA helps as well. In a nutshell, a good product manager is a combination of (a) right personality, (b) common sense, (c) solid technical skills, and (d) ability to execute. The personality should be such that the product manager should be able to persuade others to build and deliver a product per its vision. The common sense means that the product manager should be a realist and should be able to make right and timely trade-offs to deliver a sellable product on time and knows that the perfect is the enemy of good. The product manager should also know which available technologies and components to leverage to make the product a success. And finally, the ability to execute successfully on the above, that is, to balance the resources, cost, schedule, and the features of the product.

Some of the key skills one should demonstrate before becoming a product manager can be summarized as below:

- Good technical background in the field of interest
- Good understanding of the industry and trends
- Some understanding of a company's products
- Some understanding of the competitive landscape
- Basic understanding of cost accounting and finance
- Good business acumen
- Good people skills
- Good presentation skills
- Good communication and writing skills
- Good analytical skills
- Good time management skills
- Ability to handle pressure
- Strategic and big-picture thinking
- Background of working in a matrix organization

The main personality traits of a successful product manager include dealing with ambiguity, assertiveness, persuasion, negotiation, and good communication. First of all, a personality with strong communication skills is important. The product manager will need to interface with customers, sales and several functional teams, and there will be tough negotiation times. Therefore, common sense and how to put the communication skills to use are very important for the product managers. So are the presentation skills. Product managers are required to get in front of top executives and customers and put their case forward and pitch their products. They are required to express complex concepts in an easier to understand form for sales, partners, and customers. Therefore, it is advised that before stepping into the product management role, one honestly assesses those skills and sharpens them. Participating and speaking into technical events, attending customer-facing meetings, and helping out technical marketing or product management teams might be few ways to develop those early skills. Sometimes it could be good idea to step into an interim role in between such as the technical marketing engineering or system engineering role, where few years can be spent to develop the customer interface and problem solving skills. One could also decide then to rather advance in those roles versus becoming a product manager.

Finally, having business and financial acumen helps a lot. That is where MBA is usually helpful, but there are other ways to develop those skills as well. Taking few courses around fundamentals of business, cost accounting, and finance can be helpful. But more important is the ability to understand broader market and industry trends such as the total addressable markets, market share gains, revenue growth, profitability, and other financial indicators of key players in a given industry. Reading press releases and articles of a company, listening to a company's earnings calls and understanding its income statement and balance sheet as well as tracking its innovation and investment priorities help build the business acumen. Additionally, talking to the product management team frequently and asking questions, joining professional entrepreneurial groups outside of the work, and finding a good, seasoned product manager mentor can help in developing those skills and knowledge at faster pace.

Even if someone possesses most of the above attributes, it may not be easy to land on a product manager position for managing a high revenue product or a product of strategic importance for a company. The general manager would only consider very experienced and seasoned product manager to manage those products. And even if one does get the opportunity, it is a setup for failure without building appropriate real world experience first, as getting a job and performing on the job are two very different things. Therefore, the advice to the product management candidates is to find low barrier to entry spots into the product management profession to manage simpler and lighter products and then work their way up to more complex and significant products.

Products that are simple and smaller in revenue have indirect revenue contribution or are *cost centers* are viable products to start the career with. For example, in the networking industry, it may be very hard for someone to qualify for a product management job managing a multimillion-dollar data center switching product to begin with. On the other hand, one may be considered for a product management role for an entry-level product or a component of a larger product, such as optics and cables, or for products like a network management application. This strategy can help someone spend few years to build the foundation in product management profession and develop key skills without being overwhelmed. Once one has gained the basic experience in product management, has understood the product life cycle fundamentals, has handled the associated tasks and responsibilities at a smaller scale, and has known how to interact within and outside of the company with different entities, at that point one can go for handling a larger or more important product, possibly with some P&L responsibility.

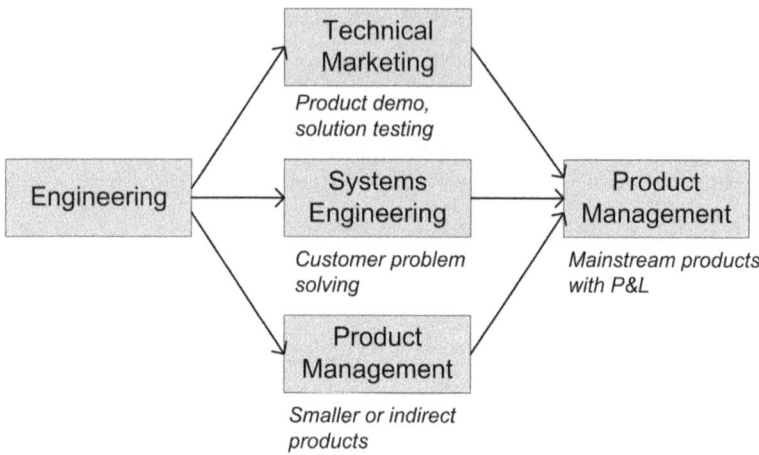

Diagram 4: Possible Transition-In Paths into Product Management

Another career transition approach is more gradual and smoother that involves changing one variable at a time. With this approach, engineers can switch tracks to the adjacent areas that utilize their technical skills much better and at the same time provide opportunities to interface with the customers and learn about the products and business aspects without formal P&L responsibility. Technical marketing can be one such area. A *technical marketing engineer* (TME) is involved with supporting the product management and sales teams with technical services, as the name sounds. This could include but not limited to activities such as testing and comparing products for competitive differentiation, managing customer demonstrations and product evaluations, managing roadshows, writing technical notes and white papers, and other such activities that are customer facing but require sound technical foundation. The content and the services produced by the TMEs are used by the product management as well as the broader sales and marketing teams. This provides the opportunity to closely work with the product management, sales, and customers and helps in learning business side of the technology, as well as building the trust and relationships that later on can help transition into product management role. Sometimes there is also a *customer solutions engineering* (CSE) or *systems engineering* (SE) team, which handles building, testing, and simulating customer deployment scenarios and performing competitive testing as well as handling the *customer proof of concept* (CPOC) for demonstrating that the products work as advertised. This also could be an alternate intermediary role.

No matter which strategy is used to step into a product management role, it must be understood that it may be a backward move in career for some time before the new foundation can be built on the new track, and the career can advance forward again. Therefore, the earlier the transition, the better it is. If it is too late in the career, it might make sense just not to do it. Also, the associated impact on work habits and personal lifestyle must be clearly understood. Not only the transition into product management role may increase the number of work hours one is used to, but may also require travel to attend customer meetings and events from time to time requiring time away from the family. Therefore, it is important to understand what changes in one's lifestyle will be required other than learning the new skill set before making the career transition. In the long run, product management can be a great and fulfilling career, never boring, to say the least.

Life after Product Management

It may be worth discussing what happens after spending major portion of one's career in product management and what the options may be for career development thereafter. The career of a product manager typically advances through a product line manager, director of product management, and then a vice president or general manager. It may continue to a C-level position such as a chief marketing officer (CMO). As we will explore in the remaining of the book, while in the role of product management, a product manager develops several outbound and business skills in the adjacent areas of product marketing, public speaking, business development, and sales enablement. At any stage, should the product manager feel motivated to transition into other areas, it is quite possible. For example, it is very possible that a product manager may find the outbound portion of its job more exciting such as public speaking, interacting with media, analysts, customer, and partners, and may want to expand into those areas. Therefore, it may consider transitioning into outbound marketing or business development role. In this case, the product manager's "soft skills" matter more than its technical foundation. On the other hand, if the product manager chooses to transition into a sales engineering type role, its technical skills may have significant weight. Yet another possibility is to step into a business strategy role such as corporate strategy or mergers and acquisitions (M&A) group. Discussing those roles is beyond the scope of this book.

While product management career can be very exciting and thrilling, it comes with its own set of challenges. Every day is a different day in the life of a product manager, bringing new challenges to solve. The degree of randomness in terms of day-to-day or week-to-week activities means that there is not much predictability in terms of the workload management. The product manager must solve whatever issues surface to keep the product moving, no matter how much time or effort it may require. The sustained pressure of getting things done timely and efficiently may not provide much *downtime* for cooling off and recharging. This is particularly true in smaller size companies where the product manager is also responsible for product marketing and additional responsibilities. Therefore, it is not surprising if some of the product managers find themselves overwhelmed in the midst of their career and want to consider transitioning out of it. This usually happens if the product manager is not being provided the opportunity for career advancement and feels stuck with the same set of activities over several years. At some point, the technical, business, and entrepreneurial skills of the product manager need to be evolved into more strategic roles.

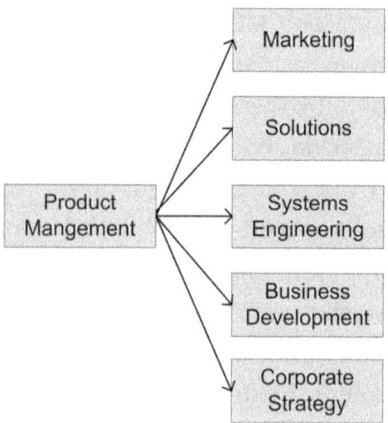

Diagram 5: Possible Transition-Out Paths from Product Management

One of the career growth possibilities that may prove to be extremely satisfying and rewarding for a product manager is to start its own business or a *start-up*. On this track, the product manager gets to satisfy its entrepreneurial spirit fully. The start-up not only requires a strong product or business idea but also a strong founding team. The product managers prove to play the pivotal role in the success of the new company since they are knowledgeable about how to turn an idea into a successful product and how to generate business out of it. They know the markets, opportunities, challenges, and customer behaviors. In this case, the product managers have proven to be successful chief executive officers (CEO). Although not all start-ups end up being successful, the ones that do result into great success story and inspiration and sometime into good financial reward. Entrepreneurship and start-ups is a vast subject, and there can be a whole book written on just the topic itself.

Key Takeaways

Product management is a demanding but exciting career. To say the least, it is not for the fainthearted. Building and launching new products and turning an idea from a piece of paper into a working product is almost a miracle. In addition, the product manager has to manage the product throughout its life and supervise it from cradle to grave. In doing so, the product manager learns about and deals with pretty much every function in the company. Only one thing is certain in the life of a product manager—no day will ever be the same.

Remember:

- A product can be a tangible or intangible item one can use in a certain way.
- Product management means defining, building, and growing products in relation to the end user requirements and for monetary purposes.
- Product marketing focuses on marketing the products versus building them.
- The product life cycle consists of a product life as it moves from cradle to grave including product planning, execution, launch, sustaining, and termination phases.

- A product manager is the CEO of a product and determines the fate of the product by making decisions about it on daily and weekly basis and throughout its life.
- A good product manager is a combination of right personality, common sense, solid technical skills, and ability to execute.
- The product manager has to engage with and direct pretty much every function within a company.
- Maintaining clear positioning among different products and have clear go-to-market strategy plays a key role in maintaining successful product lines.
- Choosing the product management as a career can be intense but satisfying. It is important to understand what changes in one's lifestyle will be required other than learning the new skill set.

CHAPTER-2

PRODUCT PLANNING

The Planning Phase

As introduced in the first chapter, the product planning is the first and the most important phase of product life cycle that the product manager must manage. The planning phase consists of several important activities including a product idea, market opportunity analysis, concept proposal, requirements definition, and even high-level product design. Planning is probably the most important phase of a product because this is the stage where a product takes shape on the canvas, and whatever shape is drawn most likely will determine final picture; the rest will be about filling in the colors. It will be fair to say about the planning phase that *plan 80 percent, execute 20 percent, and 100 percent will be OK*.

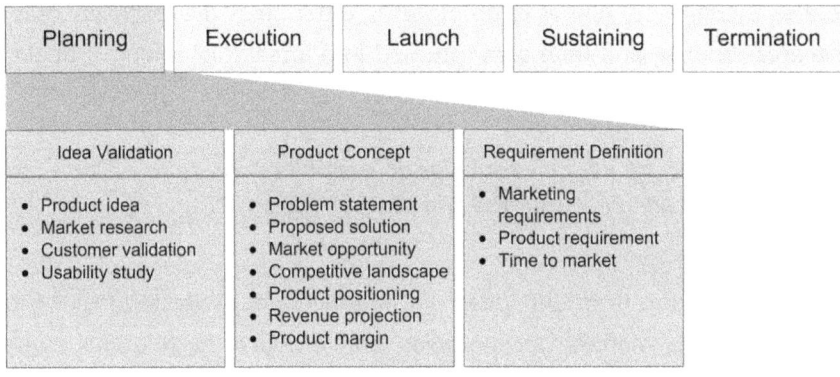

Diagram 1: The Planning Phase

Planning is the first stage where many raw thoughts enter the product manager's brain from different directions and interact in complex ways to shape a product idea. The first stage of planning is discovery and research of the useful information. The useful information is any information or data that help refine the raw product idea and channel it toward the right direction. This useful information could include market research, projections, trends, and target user feedback. Potentially, at some point, the useful information and the new product idea that the product manager is thinking will intersect, and the information gathered will either approve or negate the idea. The rest of the planning will consist of giving both the information and the idea a clearly presentable shape that could be consumed and understood by others. This is really the essence of planning, and most of the tasks discussed in this chapter are around these activities.

Product Idea Validation

If the product is going to be a brand-new product, the planning process starts with just a plain idea. An idea is usually a solution to a problem or set of problems that, if productized, a particular set of target customers will be willing to pay for. The product is meant to be designed to provide the solution to those problems. A product idea can be originated by the product manager or by anyone else who understands the industry trends and challenges and use cases to good extent. Most of the time, it is the result of a team effort. Nevertheless, it is the product manager who eventually takes the idea forward and refines it, promotes it, determines the feasibility of it as a sellable product, and then gets it turned into a real functioning product at end of the day.

> *A product idea is an unrefined solution to a problem or set of problems that, if productized, potential customers will be willing to pay for.*

Ideas usually root from the deep knowledge and understanding of the problems, needs, markets, technologies, and products or solutions already present around. They can be simply based upon repurposing or improving existing ideas or based on a totally new invention. The product manager thinks through an idea and possibly bounces it off with few colleagues, executives, or potential customers to validate it. Sometimes a formal usability study can be conducted with a set of target customers, especially when entering a new market or testing a new invention. The participating customers or end users can provide valuable feedback to the product manager if the idea could help them or not and what sort of improvements or adjustments may be required to the idea to make it more useful for them.

Sometimes this is taken a step further by developing a *mock* product that does not function much but provides a closer to realistic feel of the product or even a *bootstrap* product that only functions at bare minimum level and is yet flawed in many ways. In case of a software-only product, it could be a partially functioning application. A mock or bootstrap is usually done when it is hard to explain a complex or totally out of the box idea that may not make sense otherwise. This is done on the "seeing is believing" basis. A bootstrap is also developed to attract potential investors for taking the idea forward as serious product or when starting a new company. An advance commitment to buy from potential customers may be negotiated if the idea is a high barrier to entry and customized enough that they cannot get the product from other sources easily or if they have urgent need for such a product. However, this is rarely possible; and generally, customers tend not to make any commitments for something that does not exist yet.

As we will discuss under the execution phase in the next chapter, sometimes idea validation continues throughout the product development, and product is adjusted in the direction per the changing business or technology environment based on constant feedback loop. In this case referred to as *agile methodology*, mostly common with the software-based products, the product is developed in small increments and every time validated against the original idea and requirements. This is in contrast to the traditional *waterfall methodology* where the product idea is solidified once and the development follows in a series of sequential steps, with rare change in direction during the product development. Agile is also common with start-ups where initially it is not quite clear what the end product will be in entirety, and the product being developed the very first time may end up being quite different from the original idea.

Product Concept

After an idea has been validated with the potential customers or investors, which is an indication that they might be willing to pay for it, the product manager turns the idea into a formal *concept proposal* for presenting it to the executive leadership within the company. A concept proposal captures key information related to the idea on paper so that it can be reviewed by a *review committee* who votes on it. A review committee is usually comprised of the executive leadership and key stakeholders. If the committee approves the idea, it results into a *concept commit* (CC).

> *A concept proposal expands an idea into more refined and detailed form to be reviewed and understood by others. It explains the problem and need for a solution, the solution itself, market opportunity, product positioning, competition, and high-level business case.*

Not all companies have the concept of a concept commit; however, there is almost always a concept proposal of some sort. A concept proposal generally consists of the following sections:

Problem Statement

A problem statement describes the pain points potential customers are facing due to lack of feature functionality in existing products available in the market or due to complete absence of such a product. This defines the scope and magnitude of the *need* for a product, which must be real and not assumed. As discussed earlier, the need must have been verified with multiple sources such as potential customers to make sure that most of the customers converge on what the product manager thinks is a real problem to be solved and how it should be solved. Just assuming problems based on product manager's own knowledge or guess might result in delivering a product that no one ultimately wants to pay for, causing waste of time, money, resources, and potential impact on company's business and reputation. Part of identifying the need is to determine if the target market is ready to accept the product by the time it will be launched and not overemphasizing a problem or over-engineering a product. There are many examples of great products designed to solve problems were never the problems or that were valid problems but were still many years out. It is therefore important to solve current or near-term problems before trying to solve future problems.

Right timing to enter the market is important and so is the market acceptance. A product that is too early or too late to the market suffers from poor market acceptance. Everett Rogers addressed this issue in his book

Diffusion of Innovations, as shown with the product or technology adoption curve below. The adoption curve can be applied to the product itself for picking up the latest technology available that the product manager wants to use in the product; or it can be applied to the target customers and end users who will be using the product.

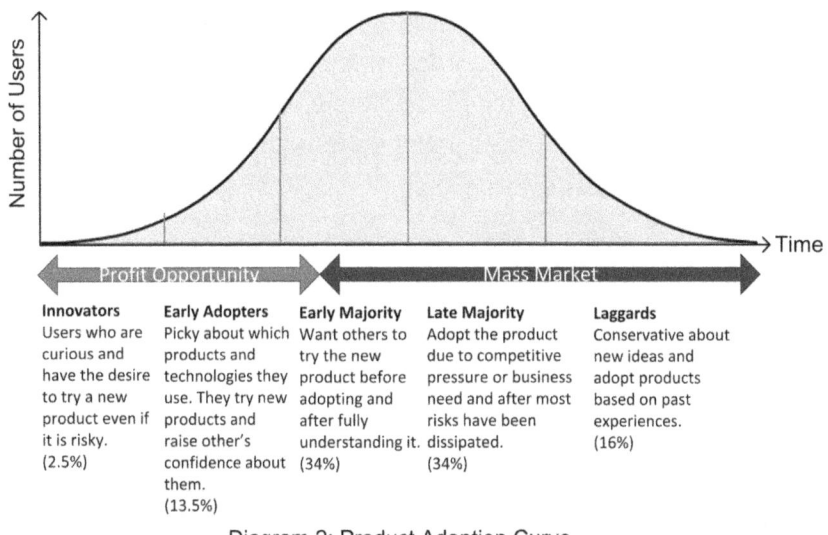

Innovators	Early Adopters	Early Majority	Late Majority	Laggards
Users who are curious and have the desire to try a new product even if it is risky. (2.5%)	Picky about which products and technologies they use. They try new products and raise other's confidence about them. (13.5%)	Want others to try the new product before adopting and after fully understanding it. (34%)	Adopt the product due to competitive pressure or business need and after most risks have been dissipated. (34%)	Conservative about new ideas and adopt products based on past experiences. (16%)

Diagram 2: Product Adoption Curve
(Everett Rogers "Diffusion of Innovations")

Specifically for the high-technology products, the product manager needs to consider what part of the new product or technology adoption curve it wants to target for entering a market and what sort of price premium it expects in exchange of providing a disruptive innovation. It may make sense sometimes to be a pioneer and lead the market to build the brand and to command higher profits, targeting the early adopters. At other times, it may make more sense to be a follower and target the early and late majority of the mass market. Regardless, the product should not overshoot the customer needs in the target market and should be absorbed by the market at the given point in time. Therefore, it is important that the product managers try to solve the problems that exist within time and scope proximity.

Throughout this book, we will use a hypothetical hi-technology company called Nubes Networks, offering advanced networking solutions for enterprise, service provider, and data center networks. We will provide realistic examples for cross-reference to understand how a hypothetical networking product gets created from scratch and how it progresses through its life as we discuss the product lifecycle in detail. This will provide readers a very good view of the product management activities in real world. Although we have chosen the networking vertical for the examples, the product management principles discussed in this book are very much applicable to any vertical within the hi-technology industry or even outside of it. Below are the examples of how the concept proposal will list the problem statements and propose a solution to solve those problems. This hypothetical product will be carried forward throughout the examples in this book.

EXAMPLE 1

Problem Statement for a Data Center Switch Product

Industry Trends:

- Internet traffic and data storage is growing exponentially.
- Bandwidth-greedy multimedia applications are consuming more bandwidth.
- More and more east-west traffic among data center resources is growing due to interactive applications, resulting in higher bandwidth utilization.
- Surge in mobility and smart devices (cell phones and tablets) are compounding the problem.
- Server virtualization is also multiplying the bandwidth usage.
- Most of the servers and storage are migrating to 10 Gigabit Ethernet putting backpressure for higher performance 40 Gigabit Ethernet switches in the core of data center.
- Ethernet is becoming the technology of choice for connectivity within the data centers due to it economies of scale, scalability, and simplicity.

Solution Need:

- Above trends require higher speeds and feeds in and out of the data centers as well as require higher bandwidth on the switches connecting resources (servers and storage arrays) together.
- The reduction in IT spending and increase in cloud based offerings put downward price pressure on capex and opex; therefore, there is need for cost effectiveness and a combination of Ethernet and the merchant silicon technologies is the way to go.

- Cost pressures and profitability needs continue to demand higher power and cooling efficiencies for lower operating costs and hence lower total cost of ownership (TCO).
- Furthermore, the rack space is getting limited to be allocated to the networking gear, and there is need for higher density in lesser footprint.

Proposed Solution

The proposed solution to the problems or needs identified above explains what type of product idea the product manager has come up with to solve those problems. The solution consists of the product definition and can be out of the box but should be possible to implement using components and technologies easily available at the time. The product should be within the desired cost constraints and likely to make money per the projections. Again, the product should not be ahead of the market, and customers are likely to pay for it, if it were made available within the predictable timeframe. Therefore, it is a good idea to validate the proposed solution with multiple potential customers before putting a concept proposal around it. Conducting some sort of contained market research, customer survey, or usability study is not a bad idea as discussed earlier.

The solution or the product definition section explains what the product will be, what functionality it will deliver, what sort of attributes and features it will have, and how it will address the customer problems. There are three basic components of the product definition that must be considered in the background, although not all of it can be fully completed by the concept stage:

- Alignment with the strategic goals and vision of the company that everyone understands and agrees with. This keeps the product aligned with the go-to-market strategy and keeps the executive leadership calm.
- Clear goals in terms of market share gain, differentiation, and customer experience. This keeps the product focused on what it is being built for.

- Leading characteristics define key features and attributes that the product needs to possess to be successful. This drives the requirements definition and execution later on.

If a product or solution is proposed considering the above factors, then it is likely to resonate with the approval committee and likely to get approved.

EXAMPLE 2

Target Solution and Product Idea:

Nubes Networks' New 10GbE Data Center Switch

- Proposing compact 1 Rack Unit (RU) high Ethernet Switch for the data center Top-of-Rack (ToR) applications.
- Product to offer 48 wire-speed ports of 1/10 Gigabit Ethernet (GbE) dual-speed ports with SFP+ interfaces on the front for Server and Storage connectivity.
- Product to offer a 48 ports 100/1000/1000 MbE RJ45 (10GBaseT) variant as part of the overall solution for connecting LAN on Motherboard (LoM) servers.
- Product to support advanced Layer 2 and Layer 3 features.
- Product to support Fiber Channel over Ethernet (FCoE) on all ports for Fiber Channel storage convergence.
- Product to support IEEE Data Center Bridging (DCBx) standards for lossless applications.
- Product to support IEEE TRILL standard.
- Product to support OpenFlow and OpenStack for Software Defined Networking (SDN).
- Product to support up to 1 Million Layer 2 and Layer 3 forwarding entries.
- Product to have a choice of 4 ports of 40 GbE for uplink connection (field upgradable).
- Product to be able to stack up to 10 units in a single stack.
- Product to support 480 Gbps stacking bandwidth.
- Product to support low latency switching, possibly < 800 nenosecond.
- Product to support front-to-back and back-to-front airflow options for hot/cold isle deployments.

- Product to support redundant (swappable) power supplies and fan tray.

- Product to target a list price per port of less than $500 in the US market.

- Product is targeted to be based on Company ABC's generation-X merchant ASIC technology.

Only important features are captured above. Detailed feature requirements will be defined later. High-level conceptual diagrams are shown as below illustrating various components and their positions. Actual product design may vary.

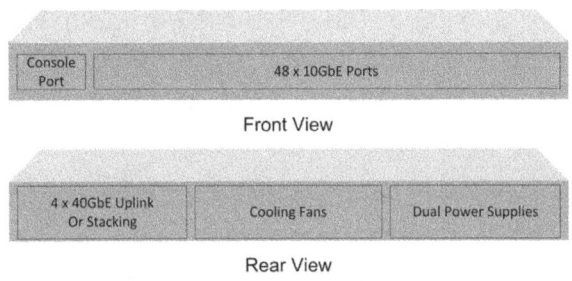

Market Opportunity

One of the key indicators of the market opportunity size is the total addressable market (TAM). The total addressable market is the theoretical maximum opportunity available to a product in terms of market size or potential revenue it could make under ideal conditions, assuming no competition. Usually, the product manager uses the market data from market research companies for firsthand assessment and calculation of the addressable market for scoping the market opportunity for the new product idea and to calculate potential revenue distribution over time. This sets the "ceiling" of the market opportunity. The ceiling helps determine how much room there could be for growth in future.

The total addressable market (TAM) is the total revenue opportunity available to a product or service in absence of any competition.

Realistically, the product revenue opportunity will be a much smaller portion of the addressable market, such as certain points of market share the new product is targeted to take from certain competitor(s). The mechanism to size up the actual or more realistic addressable market includes competition in the picture and is usually done as the next step. In this case, the addressable market is divided among all the competitors, and only a portion of the pie comes out to be the actual achievable market, although this portion could grow over time. This realistically achievable market share can be referred to as the *realizable target market* (RTM).

To calculate the realizable market, knowledge of competitor's current market share distribution is required. This information is usually available through independent market research reports. The final market opportunity is calculated by taking the overall total addressable market split into multiple competitive market shares and then calculating how much market share the new product is expected to gain year over year. The procedure is the same regardless of hardware or a software product.

EXAMPLE 3

Overall Total Addressable Market (TAM) Data Points

- As in chart below (source: Dell'Oro Research Group), it is expected that 10 GbE port shipments will increase to 62.2 million while revenues will increase to $13.2 billion in 2016.

- After 2014, 10 GbE will begin to comprise the majority of revenue in the Ethernet switch market. Fixed form factor (purpose built) switches such as ToR will comprise almost 60 percent of this opportunity.

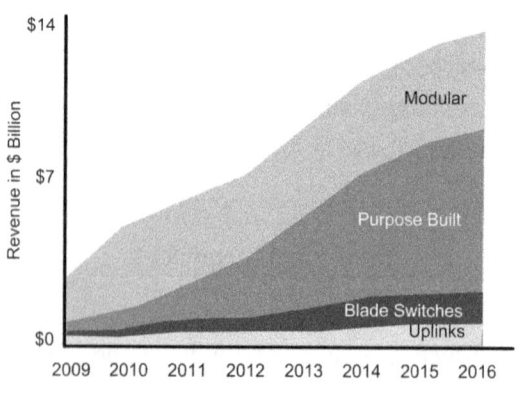

10 GbE Projected Revenues (Courtesy: Dell'Oro)

- It is anticipated that 2014 will be the year that most servers will have a 10 GBaseT LAN on Motherboard (LoM) solution. These new servers will be for the most part 10 GbE, and a significant upgrade cycle to Ethernet switches in the data center is expected to begin.

- It is expected that over the short term, blade switches will outpace the market as this is the first portion of server market to begin migrating towards 10 GbE LOM. However, over the long term, ToR switches will out-ship blade switches because majority of the server market will remain rack-mount servers.

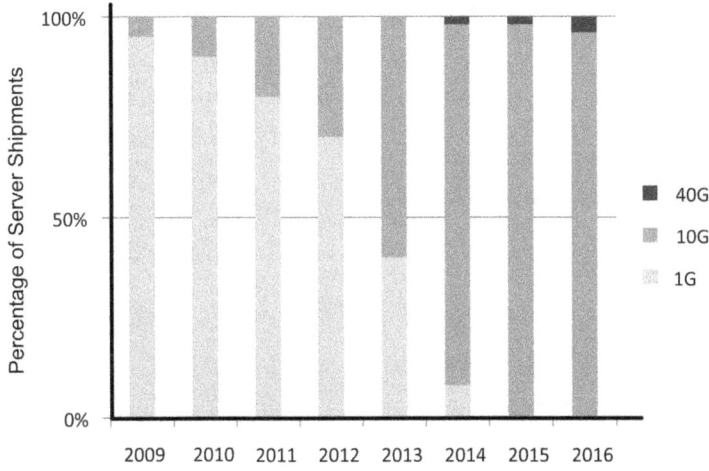

Server Attached Networking Forecast (Courtesy: Dell'Oro)

- The total data center switching market is right now about $7 billion, expected to be $14 billion in 2016. Currently, 50 percent of it is Ethernet L2/L3 switching.

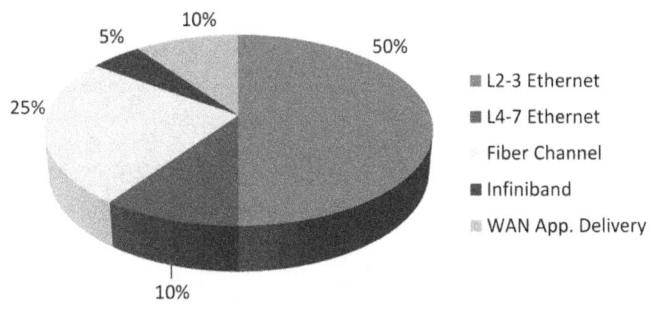

Data Center Switching Market Split

- Current data center market share split is shown below:

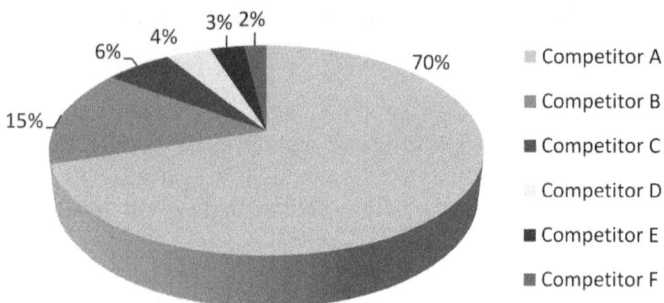

Data Center Market Share Split

Realizable Target Market (RTM)

- Per Dell'Oro, total 10 GbE TAM in 2016 (not just the data center) = $13.2 billion
- Fixed configuration 10 GbE (60 percent) in 2016 = $7.9 billion
- Data Center switching TAM in 2016 = $14 Billion
- L2-3 Ethernet Switching (assuming stays same 50%) = $7 billion
- Fixed configuration 10 GbE in Data Center (60% of the 50%) = $4.2 billion
- Therefore we can estimate that $4.2 billion out of overall $7.9 billion revenue of fixed 10 GbE L2-3 switching will be generated from the Data Center. This is our maximum TAM by 2016.
- With expected product launch at the start of 2014, assuming 1% market share gain year over year during first 3 years, targeted at competitors A and B, will result in 3% market share by 2016.
- Realizable Target Market by 2016 (3% of $4.2 billion) = $126 million

Competitive Landscape and Product Positioning

Competitive landscape explains who will be the mainstream competitors in the industry, offering similar products and solutions already in place or in the process of building such products. What those products are and how they are relatively positioned and ranked. Product *positioning* means creating the identity of the new product with reference to its attributes and pricing in the market place. Creating the right product positioning is one of the most important tasks the product manager must complete and as accurately as possible. Wrong product positioning can cause failure in the market place, even if the product is ahead in terms of time to market, because sales teams will try to sell the product in the wrong places.

> *The product positioning creates the identity of a product in the market place with reference to its attributes and price.*

Product positioning draws the circle around the product within which the product is intended to be used. Since any product cannot be best at doing all things, there needs to be clear positioning of the product. Once the positioning has been defined, competitive differentiation and value proposition of the product is defined with reference to its positioning. Differentiation provides the *silver bullets* on how the product is different from and superior to the competitive offerings. The target differentiation at the concept stage can be high level but clear enough to convey the intended winning attributes of the product. The product will be designed to deliver the intended differentiation, and it will be carried forward as the basis for creating product messaging and marketing content. We will discuss messaging in detail in later chapters.

When defining the product positioning, not only the product needs to be positioned with reference to the competitive products but also with reference to other products within the same company's portfolio or even within the same product family. This is in fact more often the case. The positioning within the product family, or *intra-product positioning*, helps clarify why there are different product offerings that apparently may look close enough but really may not be. Those products address the granular use cases as the product market reach and the install base expand and so the customer requirements. The positioning with reference to other products or product families from the same company, or *inter-product positioning*, defines how the product fits in an overall or end-to-end solution and complements it. This is usually a case that product line managers deal with. As the company's product portfolio gets richer with time, the clarity in terms of intra- and inter-product positioning becomes more important.

EXAMPLE 4

Competitive Positioning

- Primary positioning of the product will be as data center Top-of-Rack switch for 10 GbE server and storage connectivity as shown below:

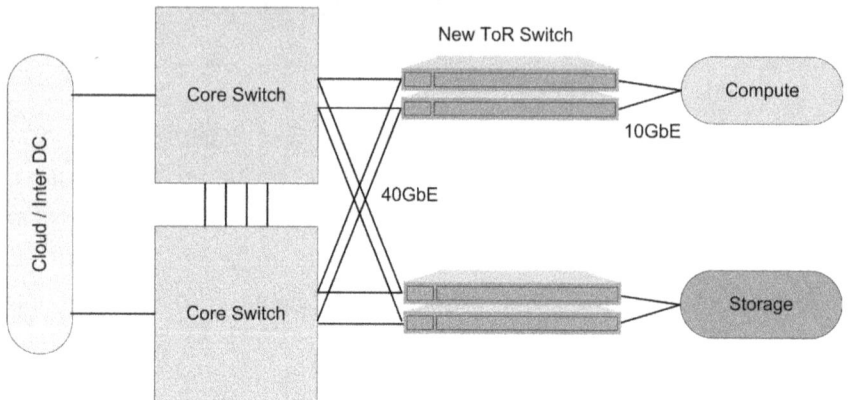

- The product will be positioned against competitor A product X and competitor B product Y.
- Primary differentiating point against the above products will be:
  - o Better price performance ratio
  - o Lower latency
  - o 480 Gbps non-blocking stacking bandwidth
  - o Open standards based 40 GbE uplink capability
  - o Better TCO
  - o Software defined networking (SDN) capability

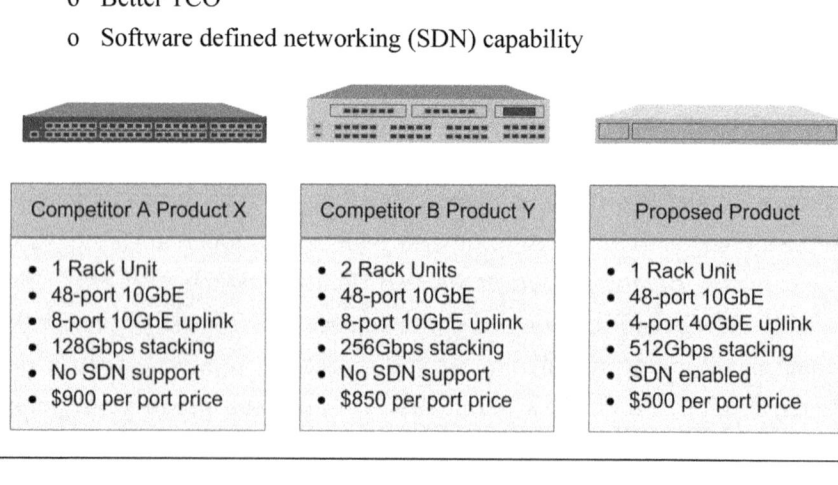

Competitor A Product X	Competitor B Product Y	Proposed Product
• 1 Rack Unit • 48-port 10GbE • 8-port 10GbE uplink • 128Gbps stacking • No SDN support • $900 per port price	• 2 Rack Units • 48-port 10GbE • 8-port 10GbE uplink • 256Gbps stacking • No SDN support • $850 per port price	• 1 Rack Unit • 48-port 10GbE • 4-port 40GbE uplink • 512Gbps stacking • SDN enabled • $500 per port price

Revenue Projection

One of the key components that the product manager needs to estimate and project at the concept stage is expected product revenue over certain period of foreseeable future. The revenue projection or *forecast* should be included in the concept proposal at a high level and to the extent the product manager can forecast at that point. Usually, it is projected for the first three years after the estimated ship date of the product. Detailed and accurate forecast is prepared later on during the execution and sustaining phases of the product life cycle based on sales guidance and actual sales trends. A revenue forecast predicts how much revenue is expected to be generated by selling the products and associated services over time, usually on quarterly basis. We will discuss more on forecasting in later chapters.

Forecasting is not an easy task, and in some companies, it can be a full-time job in itself. It is an art more than it is a science. Generally, past sales trends help a lot on predicting future sales. Since no past trends are yet available at the concept stage, it is almost impossible to forecast revenues precisely for a new product yet. Therefore, best the product manager can do is to look at some of the indirect data to build the forecast. This could include any preliminary sales and customer feedback, competitive products sales volumes estimated from their market share, and any advance customer commitments to buy. Forecast also considers seasonality of the business per fiscal year and the spending habits of the customers per industry and per demographics. In case of a new idea, the product manager takes realizable market calculated earlier and distributes it over time with expected growth quarter over quarter. This provides pretty good initial estimation for the top-line revenue growth.

> *The product forecast deals with projecting product unit volumes and revenues over a period of time, usually on quarterly basis.*

In case of hi-technology hardware products such as those used in this book for examples, it is pretty straightforward to forecast as the number of units expected to be sold. In case of hi-technology software products, unit volume could be the number of software licenses to be shipped or number of downloads, and the revenue projections are based on the revenue generated from the sales of new licenses, any renewals, and royalties. Sometimes, the software is trickier to be forecasted when it is not sold directly but is embedded with the hardware product for indirect revenue contribution or value addition.

EXAMPLE 5

Projected Volume Forecast

- The projected volume for first three years for the ToR switch is shown below in chart and tabular form:

Fiscal Quarters	Projected Unit Volume				
	10G Fiber	10G Copper	40G Uplink	Stacking	Software
1QFY1	300	150	225	270	315
2QFY1	550	300	425	510	595
3QFY1	500	250	375	450	525
4QFY1	530	280	405	486	567
1QFY2	510	270	390	468	546
2QFY2	660	380	520	624	728
3QFY2	620	350	485	582	679
4QFY2	640	350	495	594	693
1QFY3	650	360	505	606	707
2QFY3	850	390	620	744	868
3QFY3	780	300	540	648	756
4QFY3	800	340	570	684	798

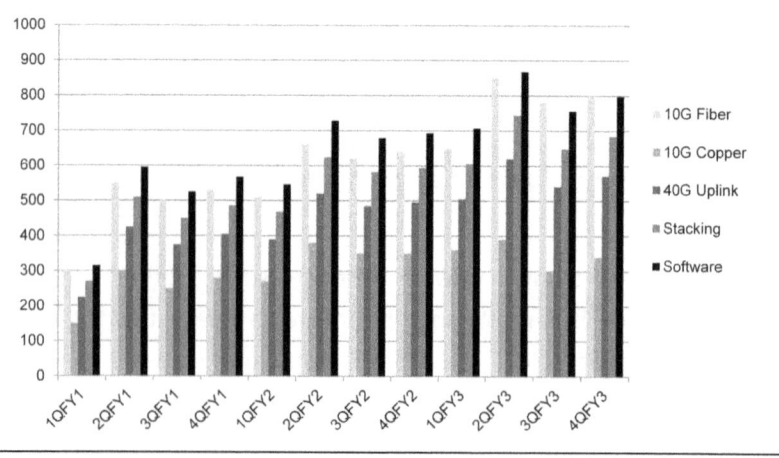

Projected Revenue Forecast

- Estimating initial unit list pricing as below:
  - o 10G Fiber Model = $20,000
  - o 10G Copper Model = $25,000
  - o 40G Uplink Module = $4,000
  - o 480Gbps Stacking Module = $2,000
  - o Layer 3 Software License = $5,000
  - o Average discount = 60 percent
- The projected revenue for first three years for the ToR switch is shown below.
- The realizable target market of $126 million is distributed over twelve fiscal quarters.
- Projected net revenue of $126,094,800 is expected at 60 percent average discount.

Fiscal Quarters	Projected Unit Volume				
	10G Fiber	10G Copper	40G Uplink	Stacking	Software
1QFY1	$2,400,000	$1,500,000	$360,000	$216,000	$630,000
2QFY1	$4,400,000	$3,000,000	$680,000	$408,000	$1,190,000
3QFY1	$4,000,000	$2,500,000	$600,000	$360,000	$1,050,000
4QFY1	$4,240,000	$2,800,000	$648,000	$388,800	$1,134,000
1QFY2	$4,080,000	$2,700,000	$624,000	$374,400	$1,092,000
2QFY2	$5,280,000	$3,800,000	$832,000	$499,200	$1,456,000
3QFY2	$4,960,000	$3,500,000	$776,000	$465,600	$1,358,000
4QFY2	$5,120,000	$3,500,000	$792,000	$475,200	$1,386,000
1QFY3	$5,200,000	$3,600,000	$808,000	$484,800	$1,414,000
2QFY3	$6,800,000	$3,900,000	$992,000	$595,200	$1,736,000
3QFY3	$6,240,000	$3,000,000	$864,000	$518,400	$1,512,000
4QFY3	$6,400,000	$3,400,000	$912,000	$547,200	$1,596,000
Sub Total	$59,120,000	$37,200,000	$8,888,000	$5,332,800	$15,554,000
Overall Revenue	$126,094,800				

- Example for 10G Fiber in 1QFY1 = 300 units x $8,000 = $2,400,000
- Where $20,000 after 60 percent discount = $8,000
- 1QFY1 = 1st Quarter of Fiscal Year 1

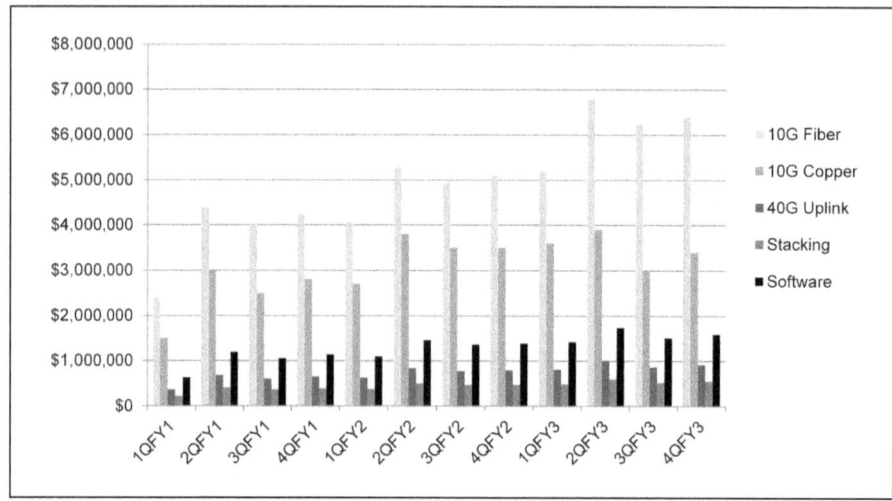

Product Margins

There are two different cost-based analyses that are needed before the execution commit stage but only one by the concept stage, a product *margin analysis*. Margin analysis provides the *profitability* picture for the new idea and helps forecast the *net income* or *profit* generated from the sales of the product. Margin analysis at the concept stage has to be estimated only as the product related costs are not fully determined yet.

> *The product margin provides the net profit picture on per product basis.*

To calculate the margin, the product manager needs to estimate the *cost of goods sold* (COGS) for the product first. COGS include the material related costs. *Raw-COGS* or *transfer-COGS* is the total cost of building a product after summing up the cost of different components and materials used in building the product, referred to as the *bill of material* (BoM). On the other hand, *burdened-COGS* is the total cost by adding any manufacturing overhead to the raw-COGS. The product manager, using help of the commodity team, can get a preliminary quote from potential suppliers or contract manufacturer (discussed later) as part of initial feasibility analysis of the idea. This homework should be done before the concept proposal. Past experience of building products also helps in estimating the COGS. A company's manufacturing overhead is usually known and can be added to get the final costs. In case of software-only products, COGS can be assumed to be zero for margin calculation purposes although practically there are some costs associated with generating *license vouchers* and other related activities.

> *The product cost of goods sold (COGS) is calculated by adding raw-COGS that is the total cost of all components and materials used per unit and the manufacturing overhead that is the cost of manufacturing per unit.*

Once the COGS have been determined, the product margin is calculated using the formula below:

$$Product\ Margin = \frac{(Product\ Average\ Sale\ Price - Product\ Burdened\ COGS)}{Product\ Average\ Sale\ Price} \times 100\%$$

The product *average sale price* (ASP) is easy to calculate. First, the product manager needs to set an expected selling price for the product that will be advertised to customers on the company's price list called the *list price* (LP). Generally, at the concept proposal stage, the product manager can only estimate what sort of price it would like to list the product at. The actual pricing analysis and final pricing is determined later on before the product launch. Initially, the product manager can use competitive products list price as a benchmark as well as manipulate the price to see how the desired margin looks like to set the initial pricing. We will discuss the pricing methodology in detail under the launch phase. Once the estimated list price has been determined per unit, the product manager also estimates the average *discount* per unit. Different markets have different discount levels, and it also varies based on the geography. We will also explore more on the regional pricing and discounting subject later on. If there is a discount that the sales teams will use for competitive bidding, then the ASP can be calculated as:

Average Sale Price (ASP) = Product List Price X (100% – % Product Discount)

Once the product margin has been calculated, it is important to compare it with the margins from other similar products, either existing or from the past. Generally speaking, in the broader scheme of product evolution, newer products should have better margins compared to older products. This is possible due to continuous improvements in the technology, supply chain, and manufacturing processes. Sometimes, addition of a new product in the product portfolio with compromised margins may be acceptable in order to complete the offering if other higher margin products are losing revenue due to the portfolio gap. In this case, the reason is more strategic than the business case based, and the aggregate product margin may be better than the individual product margins. However, mostly, the addition of a better margin product gives uplift to the existing portfolio. The new product, however, could "cannibalize" one or more existing products if the new product delivers better features and customer experience or if it is easier to sell. Sometimes the cannibalization is planned, and the product manager intentionally lets the lower margin and legacy products be cannibalized by the higher margin newer products until the demand has been shifted enough such that the older product can be gracefully terminated.

It is important to mention that the margin analysis only provides the product level profitability view including the costs from starting with the raw material up to a finished product. However, there are other costs involved in getting the product ready such as the engineers' salaries, any regulatory and safety compliance costs, and export compliance costs and so forth. Although hardware and software products can be far apart in terms of product margins based on their COGS, when it comes to the overall program costs, they are pretty comparable. In other words, developing a software-only product also has associated program costs, which could be significant. These overall program level costs are required for the full financial analysis of the project before the *execution commit* stage but not required at the *concept commit* stage. Therefore, we will explore more on this subject later.

> *The concept commit (CC) is the stage of a product life cycle process at which a concept proposal has been approved endorsing the validity of a new product idea and progress toward next stage can be made.*

EXAMPLE 6

Product Margin Analysis

- Table below calculates the product margin per product SKU (stock keeping unit) using 60 percent discount and 25 percent manufacturing overhead.
- Example calculation:
  10G Fiber Model ASP = (100%-60%) x $20,000 = 40% x $20,000 = $8,000
  Burdened COGS = $2,500 + ($2,500 x 25%) = $3,125
  Margin = $8,000 – $3,125 / $8,000 = 61%

Z-Series Top-of-Rack Switch Product SKU	Target List Price	Target ASP @ 60% Discount	Transfer COGS	Burdened CCOGS @ 25% OH	Margin @ ASP
10G Fiber Model	$20,000	$8,000	$2,500	$3,125	61%
10G Copper Model	$25,000	$10,000	$2,850	$3,563	64%
40G Uplink Module	$4,000	$1,600	$450	$563	65%
Stacking Module	$2,000	$800	$250	$313	61%
L3 Software License	$5,000	$2,000	$0	$0	100%

Once the problem statement, target solution, product idea, total addressable and realizable markets, product positioning, initial revenue projections, and product margins have been determined and captured into the concept proposal, it is ready to be presented to the executive *review committee*. The review committee goes through the proposal and clarifies any questions, assesses the market opportunity and business prospects that the product manager has put forward, and makes a decision whether to proceed forward to the next steps or drop the idea. If approved, the product manager starts preparing the detailed business case analysis and starts writing detailed product requirements, while the engineering team starts initial feasibility analysis of the approved idea. We will discuss this process in detail in the following sections and the next chapter.

Requirements Definition
The most important next step that follows the concept commit is defining and writing the detailed *product requirements*. If there is one thing that has

to be picked up as the most important task a product manager performs, it is writing clear product requirements. As the name sounds, defining product requirements means defining what is *required* to be delivered by the product. Alternatively, what is the product *required* to do in terms of feature and functionality. The requirements are communicated by the product manager to engineering and other functional teams on behalf of the target customers and users. Depending on the type of product and the development process a company chooses to implement, changing requirements during product development may be a good or a bad thing. If the product is hardware centric and the company follows a traditional waterfall development process, then defining and locking requirements upfront is important. On the other hand, if the product is software centric and the agile process is being followed, updating requirements during the product development is expected. We will discuss more on this topic in the next chapter.

> The requirements dictate what a product is required to deliver in terms of performance, features and functionality and how it will function.

In the case of waterfall methodology, it is important that the product manager does sufficient homework before writing the requirements. In this model, any change in the requirements during the execution phase is considered *requirement thrashing* and *scope creep*. One thing can be assured that engineering, and other core team members do not appreciate either frequent minor or non-frequent major changes to the requirements. Not only does constant changing of the requirements causes waste of time and resources that otherwise should have been spent already on implementing those requirement, but this also causes frustration and hurts credibility of the product manager. What it indicates is that the product manager did not listen to the customers or sales carefully, did not do enough homework before writing the requirements, or has poor understanding of the industry and the product use cases. To get the right product capabilities delivered, right requirements must be defined. In other words, to get the right answer, one needs to ask the right question. Therefore, *locking* the correct requirements early enough in the planning phase is important.

In the real world, however, there will be always some changes to the requirements as the product moves along during the execution phase; however, those changes should be fine grain in nature or otherwise called adjustments. Most of those adjustments may be the result of core team hitting any obstacles during the product design and recommending changes based on their findings and not necessarily as a result of the product manager changing the requirements. Seasoned product managers do not rely on the core team to guess and read between the lines of requirements because of their technical knowledge. Ultimately, the responsibility and the accountability of building a product as expected lies with the product manager. The product manager needs to own it and direct others throughout the process.

There are three best practice methods by which the requirements thrashing and ambiguity can be avoided, and those methods work hand in hand with each other. As the first step, capturing all requirements into a document that is referred to as a *marketing requirement document* (MRD) or a *product requirement document* (PRD); we will discuss the difference. Second, putting a review and approval process in place where all key stakeholders review the requirements document and then sign off (approve) it to endorse that they have read and understood the requirements clearly. The requirements document should be version controlled to track the changes and to avoid confusion. Third, a mechanism to document the responses to the requirements should be documented indicating what can or cannot be delivered per the requested items. We will discuss these topics in detail next.

In case of the software products, products in intense competitive environments, or if the agile development process is being followed, constant adjustments in the requirements is in fact expected. In this case, the product development happens in short increments or *sprints*. At every increment, the product development is paused and cross-checked against the changing business environment to make sure that the product is still relevant to the market and still desired by the customers. Any adjustments or updates that may be needed to the product requirements are made, and in this case, the product requirements document is a living and evolving document throughout the product execution phase. In this case, what matters more is that the right product is developed, the one that meets the customer needs and can sell versus a product that may end up on the shelf, even if it was developed per the original requirements.

Marketing and Product Requirements Documents

As mentioned earlier, the marketing requirements document (MRD) or the product requirements document (PRD) is the starting point of requirements definition. Generally, there is a subtle difference between a marketing requirements document and a product requirements document. A marketing requirements document usually defines only high-level and market-level requirements that are needed to create a new market or to be successful in a certain market. Those requirements are generally not very concrete to be actionable in a specific way. For example:

> We must build a switch product to complement our server play
> in the data center market.

The marketing requirements document could include information such as total addressable market, market share information, business objectives, winning strategy, and business case, etc. It is more of a strategic document focused on the *go-to-market* (GTM) aspects. We will discuss GTM strategy in detail toward the end of this book.

A marketing requirements document (MRD) defines market requirements that are needed to create a new market or to be successful in certain market as well as associated go-to-market strategy.

The marketing requirements document usually lives long term and spans multiple product life cycles. Chunks of market requirements get implemented in multiple phases in the form of one or more products and solutions. The requirements in the marketing requirements document do not mandate necessarily building a product to address certain market needs. A *build* versus *buy* versus *partner* analysis can be considered before building a product. A marketing requirements document can be originated by someone in the corporate business strategy team, by the solutions marketing team, or by

---

the product manager. Usually, the product manager combines the marketing requirements document and the product requirements document into a single document. A product requirements document translates higher-level strategic and marketing requirements into more concrete and actionable tactical requirements that can be translated into a product in terms of a feature and functionality, given that only building a new product is the solution. Example of a requirement in the product requirements document is:

> Build a 1RU wide Ethernet switch that must have 48-ports of wire-speed 10 GbE SFP+ type on the front panel, for server and storage connectivity.

A product requirements document is always originated by the product manager. In this section, we will explore the most common approach taken with a combined marketing requirements document and product requirements document, and it is still referred to as a product requirements document due to its main focus on the product and not on the market requirements.

> A product requirements document (PRD) defines the requirements for creating a new product and what capabilities the product will possess.

The "hybrid" product requirements document takes the portion of the overall marketing requirements that would apply to the new product and explains the market rationale and opportunity for building the product. It goes over the product idea, target solution, expected time frame to market, and initial volume projections. The heart of the product requirements document consists of the detailed product requirements explaining what is required in terms of product attributes such as form factor, physical look and feel, feature and functionality, industry standards compliance, packaging, testing, and anything else that the product manager wants to be delivered. In a start-up environment, a product requirements document could start with notes on a napkin or a snapshot from the whiteboard discussion and few pages documenting them; while in an established company, it could consist of a detailed and thorough document.

Sections of a Product Requirements Document

Executive Summary

The executive summary, as the name sounds, summarizes the point of view presented in the product requirements document and the objectives of implementing it before forcing someone to read the details. This is the start of the product requirements document and is usually a couple of paragraphs.

EXAMPLE 7

Executive Summary

Overall Ethernet/IP market is growing, and Ethernet is particularly being the medium of choice for the campus, data center, and service provider connectivity options. Demand for cost-effective, high-speed Ethernet interfaces continues to grow. This promises a huge opportunity for us being a networking company.

This PRD goes over the key requirements for building a new 10 Gigabit Ethernet Top-of-Rack switch to take advantage of the above opportunity in the data center space. It goes over the target product positioning, competitive analysis, and expected volume and revenue projections.

Market Opportunity Overview

This is the section of the product requirements document that is actually a squeezed down marketing requirements document. This is the section that explains overall market opportunity and rationale, otherwise in a separate marketing requirements document. It explains what problems are there to be solved, what is the associated market opportunity if the product was built to solve those problems. This section bridges the gap between deep down requirements that will follow and the overall rationale for asking those requirements. Preliminary analysis done during the concept proposal stage such as total and realizable market analyses, problem statement, and proposed solution can be leveraged in this section with details added. This section should be written in a way that the readers, who belong to multiple functional groups and may not have broader market or business knowledge, can easily understand and digest the information provided.

EXAMPLE 8

Market Analysis and Opportunity

Per Dell'Oro reports, projected Ethernet (L2+L3) port shipments will grow to 383 million, and revenues will grow to $27.5 billion in 2016. As in the chart below, it is expected that 10 GbE will be major portion of the Ethernet market with port shipments will increase to 62.2 million while revenues will increase to $13.2 billion in 2016. After 2014, 10 GbE will begin to comprise majority of the revenue in the Ethernet Switch market. Fixed form factor (purpose built) switches such as ToR will comprise almost 60 percent of this opportunity. The total data center switching market is right now about $7 billion, expected to be $14 billion in 2016. Currently, 50 percent of it is Ethernet L2/L3 switching so significant 10 GbE opportunity lies within the data center.

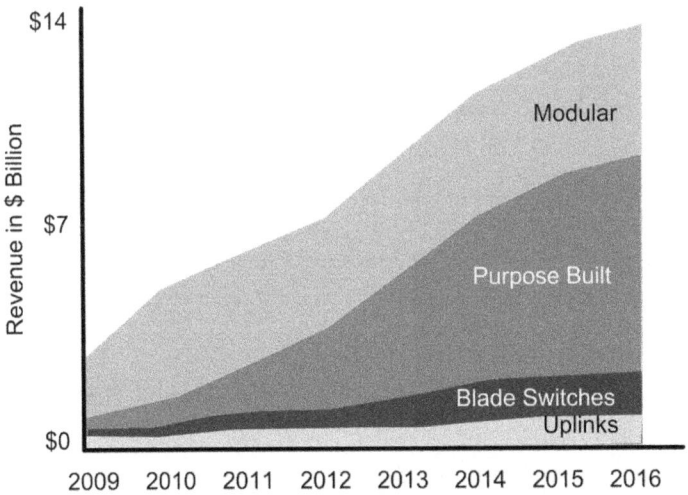

10 GbE Projected Revenues (Courtesy: Dell'Oro)

It is anticipated that 2014 will be the year that most servers will have a 10GBaseT LAN on Motherboard (LoM) solution. These new servers will for the most part be 10 GbE, and a significant upgrade cycle to Ethernet switches in the data center is expected to begin. It is expected that over the short term, blade switches will outpace the market as this is the first portion of server market to begin migrating toward 10 GbE LOM. However, over the long term, ToR switches will out-ship blade switches because the majority of the server market will remain rack-mount servers. In the next couple of years, most large enterprises will likely upgrade to 10 GbE for server access through a mix of connectivity options (all server vendors will not have 10 GbE LOM, and copper will remain dominated by SFP+ direct attach). It is therefore an opportunity for us to capitalize on this 10 GbE opportunity within the data centers for server and storage connectivity by providing competitive 10 GbE price per port with the new proposed product in this PRD.

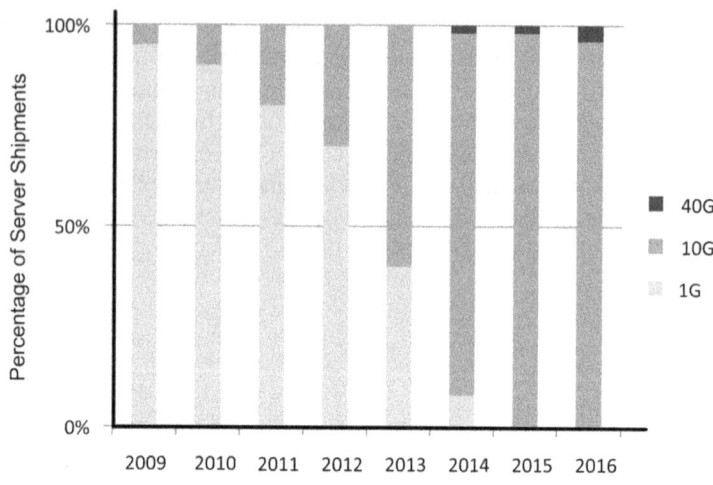

Server Attached Networking Forecast (Courtesy: Dell'Oro)

Product Overview

This is the section in a product requirements document that explains what the product will be and possibly how it will look like. What will be the high-level physical attributes of the product, and how and where it will be positioned. The reason for introducing the product in the background of competitive landscape is highlighted. This section could also go into estimated price of the product, marketing orderable part numbers, and other relevant details to help broader teams understand as much as possible about the anticipated product. The product overview does not have to be exactly how the actual product will end up being. The idea is to define the framework and have the actual product be as close to it as possible.

EXAMPLE 9

Product Overview

The product being requested in this PRD is a 1 Rack Unit (RU) high Ethernet Switch for the data center Top-of-Rack (ToR) applications. Diagrams below show the high-level product concept (that is subject to change) and the target use case.

Product will offer 48 ports of 1/10 Gigabit Ethernet (GbE) dual-speed ports with SFP+ interfaces on the front for server and storage connectivity at wire speed. It also has console and management ports on the front panel. On the rear, switch will have a swappable uplink module or stacking module, swappable redundant power supplies and fans. Exact location of the above items will be determined during the design. Details are discussed under the requirements section.

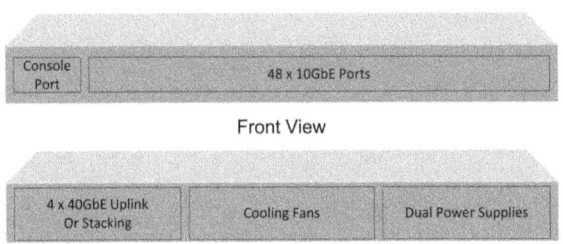

Primary positioning of the product will be as the data center Top-of-Rack switch for 10 GbE server and storage connectivity.

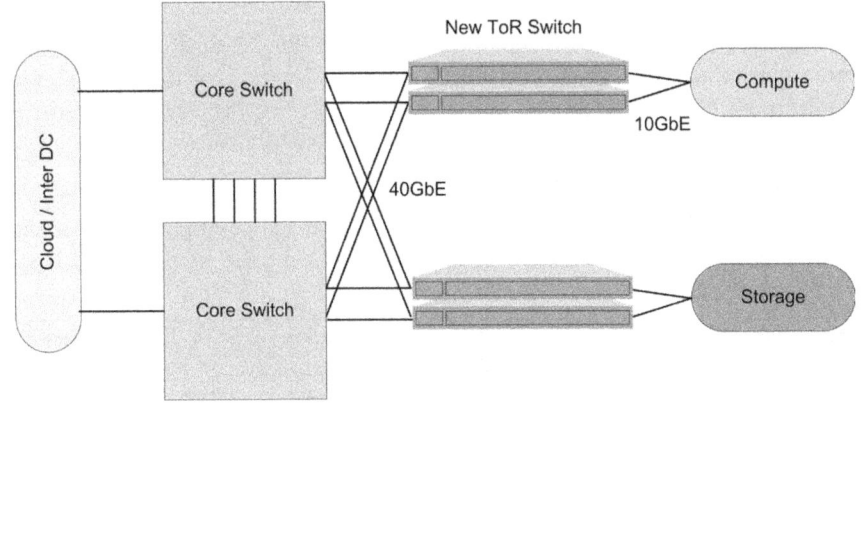

Competition

There are multiple Top-of-Rack products available in the market; however, the proposed product will be positioned primarily against competitor A product X and competitor B product Y. Those two competitors control the largest market share and there are more alternatives being sought out against those products. This provides a good market share gain opportunity, which has been assessed outside of this PRD as business case. Primary differentiating point against the above products will be:

o Better price performance ratio
o Lower latency
o 480 Gbps non-blocking stacking bandwidth
o 40 GbE capability
o Better TCO
o Software defined networking (SDN) support

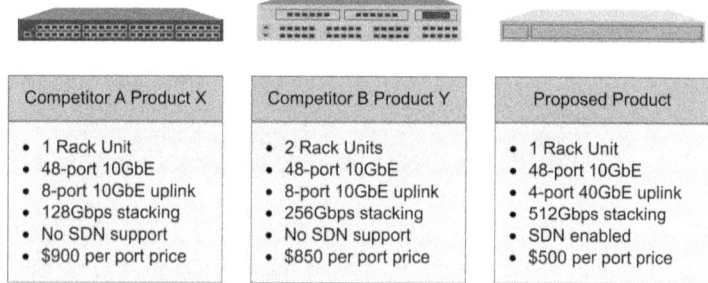

Competitor A Product X	Competitor B Product Y	Proposed Product
• 1 Rack Unit • 48-port 10GbE • 8-port 10GbE uplink • 128Gbps stacking • No SDN support • $900 per port price	• 2 Rack Units • 48-port 10GbE • 8-port 10GbE uplink • 256Gbps stacking • No SDN support • $850 per port price	• 1 Rack Unit • 48-port 10GbE • 4-port 40GbE uplink • 512Gbps stacking • SDN enabled • $500 per port price

Product Naming and Target Pricing

The product will be a new family referred to as the Z-Series. The marketing part numbers and target initial pricing (subject to change) is listed here. All part numbers are new, and there are no replacements to be made. All prices are US list prices, and other regional prices will be derived from those at the time of Open-Books.

Below are the initial marketing part numbers, SKUs, and descriptions. All pricing is preliminary and is subject to change when detailed price proposal package is prepared outside of the PRD.

Part Number	SKU	Description	List Price
101	TOR-10G-48F	48-port 10GbE SFP+ ToR Switch	$20,000
102	TOR-10G-48C	48-port 10GbE RJ45 ToR Switch	$25,000
103	TOR-UL-40G-4	4-port 40GbE Uplink Module	$4,000
104	TOR-STACK-480	480Gbps Stacking Module	$2,000
105	TOR-L3-LIC	Layer 3 Feature License	$5,000

Expected Time to Market

This section in the product requirements document sets expectations in terms of what the product manager expects in terms of the *time to market* (TTM) for the product delivery, starting with the requirements up until *general availability* (GA) of the product. The product manager should have a clear idea when the product should get ready and start shipping as this may greatly affect the business outcome. A late product, even the right one, may lose its effectiveness in terms of surprise factor, market lead, and competitive advantage if the competition catches up or overtakes.

However, being first to market is not always good. Sometimes, it makes sense to be a follower rather than a leader and learn from the competition's mistakes.

The expected time to market is more of a request to the core team and acts as a guideline for the delivery time frame. The actual delivery schedule of the product is prepared by the core team's program manager or by a project manager. This is done after the product requirements document has been reviewed by the representatives of all functional groups (the core team), initial product feasibility has been completed, and the resources have been allocated. The resource, costs, and schedule calculations must be finalized before entering the project commit stage that marks the start of execution phase as discussed in the next chapter. In case of agile method, the time to market may define multiple shipment dates for different iterations of the product, such as multiple software releases.

EXAMPLE 10

Expected Time to Market

The time to market is the most important factor for this program. The requested product is expected to go out for Beta Trials by the end of next calendar year with and expected GA around one quarter later. Actual schedule will be delivered by the NPI program manager.

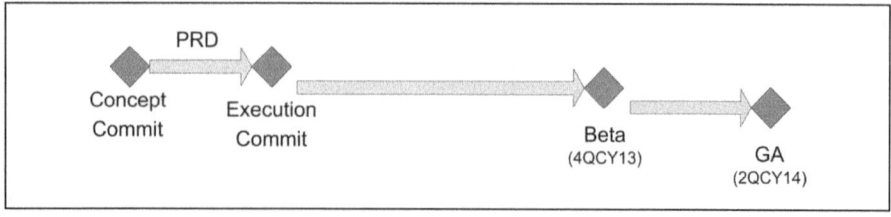

Marketing Forecast

The forecast section in the product requirements document is meant for the operations team to start having the supply chain negotiations and to start the real costs (COGS) estimates, as discussed in the next chapter. It is also used by the finance team and the core team's program manager to work on the financial *return on investment* (ROI) analysis of the program to calculate the overall program costs. The marketing forecast prepared earlier during the concept proposal can be reused here. Usually, the forecast for the first three years is needed for proper planning.

EXAMPLE 11

Marketing Forecast

This forecast is expected to vary based upon the GA time frame shift. The forecast provided is for the initial guidance purpose only. The final forecast will be provided after the target cost and pricing has been finalized and a schedule has been locked.

Fiscal Quarters	Projected Unit Volume				
	10G Fiber	10G Copper	40G Uplink	Stacking	Software
1QFY1	300	150	225	270	315
2QFY1	550	300	425	510	595
3QFY1	500	250	375	450	525
4QFY1	530	280	405	486	567
1QFY2	510	270	390	468	546
2QFY2	660	380	520	624	728
3QFY2	620	350	485	582	679
4QFY2	640	350	495	594	693
1QFY3	650	360	505	606	707
2QFY3	850	390	620	744	868
3QFY3	780	300	540	648	756
4QFY3	800	340	570	684	798

Product Requirements

This is the section that comprises the heart of product requirements document and explains what the product manager wants delivered on the target customer's behalf. The requirements section must be clear and concise so that there is no ambiguity and there can be only one meaning derived out of a given requirement. Writing clear and accurate requirements is an art that comes with experience and demands knowledge about the industry, technology, products, competitive landscape, and customer environments. Requirements can be *descriptive* or *prescriptive*, depending on who are the recipients of the requirements, how much experience the team has building similar products in the past, and how much wiggle room the product manager wants to provide to the team to use their own discretion during the product design. Example of a descriptive style requirement is as follows:

> *The switch will support 48 ports of 10GbE SFP+ type and 4 ports of 40GbE QSFP+.*

The example of prescriptive style requirement is as follows:

> *The switch will have 48 ports of 10GbE SFP+ type located on the front panel split across two rows. The 4 ports of 40GbE QSFP+ type will be supported on the rear side of the switch left of the power supply units all in a single row.*

Requirement Modularity and Traceability

There are many ways of documenting requirements; however, organizing requirements in a *modular* way that is easy to understand, and is traceable, is important. Modularity means that the requirements are segmented into multiple sections and are identified with clear titles. Traceability is accomplished by numbering each requirement and by identifying each requirement section with a *Requirement ID*. Traceability is important because different documents such as product design document, engineering functional specifications, and test plans can link back to the specific requirements in the product requirements document. Traceability ensures that no requirements are missed from the implementation or testing processes.

This backward linking or traceability is also important to demonstrate for the quality process certifications and audits such as ISO (International Standards Organization) certification that many hi-technology companies desire.

In addition to the traceability, it is also important to prioritize the requirements in relations to each other. This helps the core team understand, as recipient of the requirements, that where they need to focus first or focus more. This particularly helps under the situation when resources are constrained and budgets are tight. In case of software products such as new software features, the prioritization process is even more important and therefore starts early enough even before a product requirements document is written. These methods has been implemented and tested by the author and work well as shown in the examples that follow.

Requirement Prioritization

Before the requirements must be communicated to the core team, they must be filtered and prioritized. Filtering means that the product manager should, first of all, reject the requirements that are definitely not going to be catered to. That is, first of all, it is important to determine what is *not* being done. There could be requirements fed by the sales team or others that may be far from overall strategy alignment, product goals, or target markets. Those could be filtered up front. This leaves to prioritize among what *is* going to be implemented. The next most important step is to prioritize the requirements in the order of what is most important to be delivered first. "What is important" could be very subjective, but there are some overall guidelines for how to prioritize and rank the requirements. Requirements should be prioritized along multiple scales and not just what the product manager thinks is important unilaterally. For example, those scales could be customer rank (external), customer importance (external), customer experience (external), business impact (internal), and funding and resource availability (internal).

Customer rank could mean how important the customer is for the company, such as a global critical account who gives repeat business to the company verses a small customer. Customer important could mean how important the requirement is to a customer, especially when the same customer has asked for multiple requirements. Customer experience could mean many things. For example, when customer has a choice to choose between two products that provide similar functionality, it will likely choose the one that is easier to use. Therefore, consideration to customer experience and usability of the product is increasingly becoming important. Business impact means whether implementing a requirement will result in generating revenue or whether not doing so will result in loss of revenue or not. Finally, the funding and resource availability, which equates to the effort required to implement a requirement, means how expensive or complicated a requirement could be to implement. Sometimes it makes sense to implement four different requirements that would generate more business and use less effort to implement instead of implementing one heavyweight requirement.

Requirements	Customer Rank	Customer Importance	Customer Experience	Business Impact	Effort	Overall Priority
Requirement 1	2	1	3	2	4	12
Requirement 2	1	1	1	1	3	7
Requirement 3	5	2	1	5	5	18
Requirement 4	4	1	2	4	2	13
Requirement 5	3	1	4	3	1	12

Diagram 3: Requirement Prioritization (Lower = Higher)

In case of software features and productizing that is usually in the form of new software releases, the product manager starts gathering the new feature requests from customers and sales teams through a process of *feature request* (FR). This process is only needed if a feature idea is being submitted by someone other than the product manager himself. The feature request process can apply to both the hardware and the software products although it is rare for a hardware idea submission. The product manager can always originate the idea of a new software product or feature and take it through the concept commit. Unless the software product idea is big enough, the concept proposal usually consists of collection of several software features bundled together. A feature request outlines the overall business opportunity, potential customer name(s), description of the feature requested, and the time frame by which it is requested.

> *A feature request (FR) communication process enables customers and sales to request a particular product feature or functionality from the product manager that is important to them.*

The product manager collects all feature requests submitted in a given time frame window, reviews them, and *accepts* or *rejects* the requests based on alignment with the company strategy, resource availability, business case, and other criteria. Out of all accepted feature requests, the product manager then ranks them in terms of relative priority based on the business impact and urgency to market or using a prioritization matrix discussed earlier. It is usually hard to judge the direct revenue impact of software features and prioritize them purely based on that basis, unless it is an independently sellable software product. We will explore more of this under the business case discussions later in the book. In case of a software product supporting a hardware product, indirect revenue contribution and enablement of sales for the hardware product, as well as the strategic impact of the feature should be considered, and so should be the *lost opportunity cost* due to not having the feature, as discussed earlier. Not having a feature may prevent a product from being deployed in certain use cases, even if it has several other advantaged and hence the product sales may be impacted.

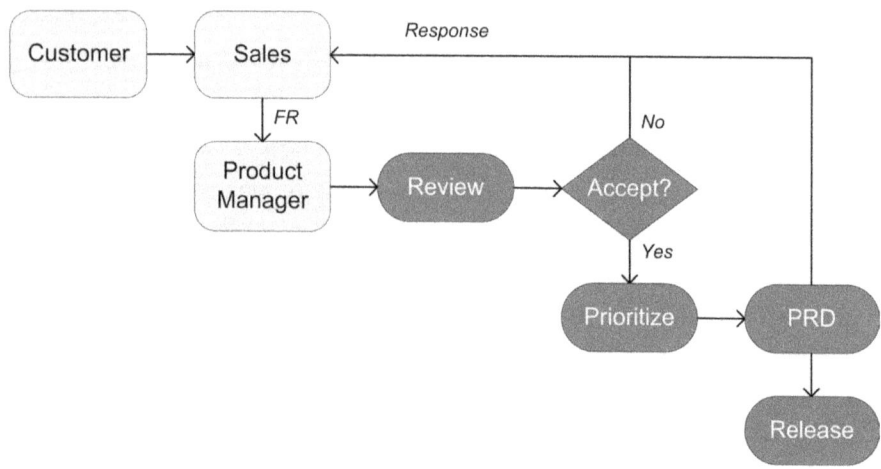

Diagram 4: Feature Request Process

Documenting Requirements

In addition to the prioritization among the product requirements, one of the best practices that has been pioneered and tested by the author is adding a *deviation* field to each requirement in the product requirements document. As we will explore more in the next chapter, the core team responds to each requirement in the product requirements document after review and assessment, whether the requirement can be actually implemented as asked or not. And if the appropriate set of capabilities, resources, and technologies are available to deliver those or not. The core team responds to the product requirements document through a separate document called a *product design document* (PDD). This is, however, a complicated process, that requires requirement traceability, adds delay, and mostly it is not written at all, causing disconnects and confusion during the product execution phase.

Adding a deviation field in the product requirements document itself captures any responses or intended deviations from the requirements, closes any disconnects, and removes any ambiguity. The net result is a clearly documented agreement on what will be actually delivered. There is no room for confusion later on. The deviations can be captured by the product manager at the time of product requirements document is reviewed, or they can be filled in by the core team members directly before the requirements are locked. In this case, product requirements document approval means that both the product manager and the core team members are agreeing on the adjustments made. The product manager can ensure the validity of deviations by getting back to the customers or sales team. The deviation field becomes particularly useful when the agile methodology is used, since incremental and frequent adjustments to the product requirements may need to be made and tracked during the product development. In this case, every time a new revision of the document needs to be signed off after reviewing the changes only.

It is a wrong practice to change the requirements themselves to indicate what the core team actually can deliver. The original *ask* by the product manager should always remain in the product requirements document as it was requested on behalf of the customers. This helps the product manager later on in situations where customers may dispute on what they had asked for versus what was actually delivered to them, or if the core team may move the burden on product manager for not communicating any requirement clearly if product misses to implement any. The deviation field should be used to capture anything different from the original requirement.

EXAMPLE 12

Product Requirements

The following table explains how the requirements are prioritized in this PRD:

Priority	Description
Mandatory	Requirement is a must deliver overall* and will be considered a GA stopper
Highly Desirable	Requirement is important but will not be a GA stopper and has business impact
Desirable	Requirement is nice to have and will not be a GA stopper. Can be on roadmap
Deferred	Requirement is deferred at this point to a future phase
Dropped	Requirement is dropped for now for indefinite time
No Change	Requirement is in line with implementations on previous products
* Unless otherwise deviations are clearly captured and approved by PLM as part of this PRD	

Detailed requirements begin below:

Requirement ID	TOR.V1.R1
Priority	Mandatory
Title	10 GbE Top-of-Rack Switch Fiber Model
Category	Hardware
Products	Z-Series
Description	A new Top-of-Rack switch must be designed and built using merchant silicon technology: a) Switch must support 48-port of 10 GbE SFP+ in 1RU height on the front. b) Every port must be dual-speed 1/10 GbE and wire speed. c) Switch must support non-blocking front port-to-port switching. d) Switch must support RJ45 management port, console port, and 2 USB ports on the front. e) Switch must support a hot-pluggable 4-port 40 GbE QSFP+ uplink module pluggable on the rear of the switch. f) Switch must support stacking of up to 8 switched at minimum, 10 units as a goal through a pluggable stacking module on the rear. g) Switch must support minimum 480 Gbps of stacking bandwidth, 512 Gbps is desirable.

	h) Switch must support less than 1 microsecond port-to-port latency. Goal is 500 nenosecond. i) Switch must support redundant, load-sharing. Hot swappable AC and DC power supplies. j) Switch must support front-to-rear airflow and hot-swappable fan module.
Deviation	[HW Engineering] Stacking of 10 units will be a stretch using the technology and cost targets and cannot be guaranteed.

Requirement ID	TOR.V1.R5
Priority	Mandatory
Title	Hardware Dependent Features
Category	Hardware
Products	Z-Series
Description	Following hardware dependent features are required when selecting the merchant ASIC: a) Switch must support advanced layer2 and layer3 features with up to 512K L2/L3 entries scale. 1M is desirable. b) Switch must support Fiber Channel over Ethernet (FCoE) on every port. c) Switch must support IEEE Data Center Bridging (DCBx) standards including PFC, FIPS, ETS, and QCN. d) Switch must support IEEE TRILL. e) Switch must support VXLAN.
Deviation	[SW Engineering] current merchant ASIC does not support VXLAN functionality and therefore this requirement cannot be met.

Key Takeaways

The product planning phase is the most critical phase, as it should be clear by now. Anything done well during the planning phase will pay off during the other phases of the product life cycle, and anything done poorly during the planning phase will result in problems and surprises later. Therefore, if any phase deserves close attention of the product manager, it is the planning phase. In the next chapter, we will explore how a well-thought-out and well-planned product actually starts taking shape.

Remember:

- Product planning is the first and the most important phase of product life cycle. Plan 80 percent, execute 20 percent.

- The planning process of a product starts with a plain idea. A concept proposal expands an idea into refined and detailed form to be shared with others.
- The product idea is an unrefined solution to a problem or set of problems that, if productized, potential customers will be willing to pay for.
- The concept proposal includes the problem statement, proposed solution, total addressable and realizable market, competitive positioning, projected volume and revenue, and expected profit margin.
- Total addressable market is the total revenue opportunity available to a product or service in absence of any competition verses a realizable market, which is realistic market share that the product can gain over time.
- When defining the product positioning, not only the product needs to be positioned with reference to the competitive products but also with reference to other products within the product portfolio.
- A product requirement document translates higher-level market requirements into more concrete and actionable tactical requirements that can be implemented as a product in terms of a feature and functionality.
- Not having a feature may prevent a product from being deployed in certain use cases, resulting in lost opportunity cost.
- A revenue forecast predicts how much revenue is expected to be generated by selling the product over time, usually on quarterly basis.
- The product margin provides the net profit picture on per product basis. Generally speaking, newer products should have better margins compared to the older product.

CHAPTER-3

PRODUCT EXECUTION

The Execution Phase

Once the product concept proposal has been approved by the review committee, that is, the *concept commit* (CC) has occurred and the product requirements have been written by the product manager, the next step is to get the product requirements document signed off by the key stakeholders. This sign-off is important to reach an agreement on requirements so that later on there is no confusion regarding if the requirements were not communicated or not understood clearly. Usually, this must be completed before a milestone referred to as an *execution commit* (EC). The execution commit follows the concept commit whereas a product moves from a bare concept to a concrete project to execute upon by all parties involved, and the required resources and the funding has been secured.

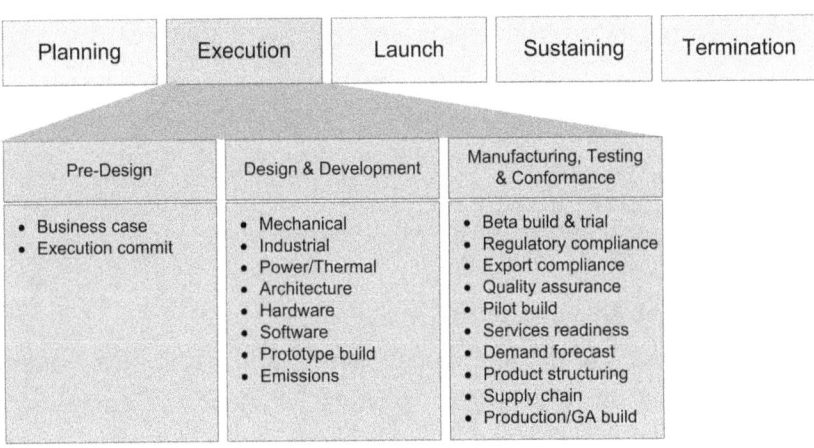

Diagram 1: The Execution Phase

Many companies implement what is called a *product life cycle* (PLC) process to manage the movement of a product throughout its life. Each milestone in the product life cycle process is referred to as a *stage*, a *phase*, or a *gate*. A stage has well-defined entry and exit criteria that must be met in order to *exit* that stage and move forward toward the next stage in the life of the product. The product life cycle process ensures all the tasks and deliverables that must be completed at every stage and identifies owners for those tasks per the overall *total quality management* (TQM) process of the company. It also provides the templates to be used to document those deliverables and how the documents will be approved. This is also important from the ISO certification and audits perspective. The PLC stages are usually denoted by numbers such as stage 0 (S0), stage 1 (S1), and so forth, as shown below. Alternatively, some companies call them phase 0, phase 1, and so forth.

> The product life cycle (PLC) process manages the movement of a product throughout its life in a well-defined way by identifying the entry and exit criteria, key deliverables, and owners at every stage of the life.

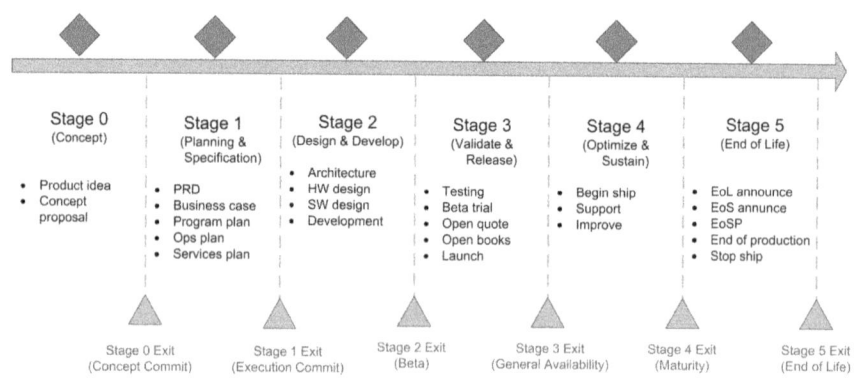

Diagram 2: Typical PLC Process Stages

Development Methodologies

Before we discuss the product execution phase in detail, it is important to discuss some of the generic product development concepts and methodologies that are used commonly today and a product manager should be familiar with. There are two major methodologies used in the industry for the product design and development. The first one is the rather traditional *waterfall methodology*. The waterfall method consists of sequential development steps, one followed by the other and one being completed before the next one can start. The typical steps include requirements, design, development, verification, and maintenance stages. With the waterfall method, it is assumed that all requirements have been gathered, understood, locked, and communicated before starting the product development. In many cases, this results into issues. For example, in case of a software product, it is difficult to tell the engineering team everything that needs to be present in the software before the software has reached certain degree of completion and can be run and tried. Only then loopholes and deficiencies will be discovered. Similarly, the design and architecture must be completely locked before proceeding to the development and so forth.

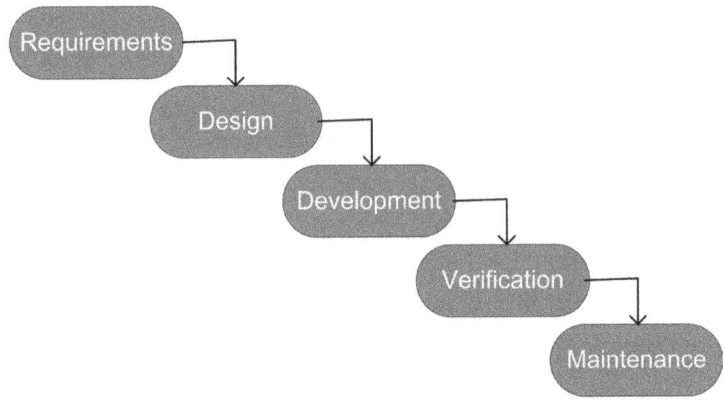

Diagram 3: Waterfall Methodology

The waterfall method suits the manufacturing environments or hardware products better where late changes to the product can be extremely difficult and cost prohibitive. In the waterfall model, the engineering team has only one shot at getting it right in all aspects. In this method, the major upfront design approach is used to minimize the risk for expensive changes later on. As we will explore later, we will learn that a defect is much more expensive to fix later in the product development stage than it is sooner. Major emphasis on the architecture and design is put upfront. It is quite possible that using this method, the engineering team will deliver a product per what was asked for, but it is also possible that meanwhile the business realities have shifted and hence the requirements as well, resulting in a completed product that no customer wants to pay for. However, this risk is lower for the hardware products typically. It is worth noting that this method could still be used for those software products that have stable requirements and high barrier to entry ideas, less likely to change.

> *The waterfall method consists of sequential development steps, one followed by the other and one being completed before the next one can start.*

The second one called the *agile methodology* is a rather modern and alternative product development methodology that is mostly used for developing software products in highly competitive markets. Agile was developed to handle large-scale software product development projects, which should not be developed like a typical hardware-based product. Typical hardware product is developed in sequential increments such that one part should be finished before another one can be started, such as in the waterfall method. As we have discussed, under such approach, all requirements must be gathered and communicated by the product manager before the execution commit and locked into the product requirements document. Then the product architecture and design is completed, and only then the real product development can start. Due to the serial nature of this development approach, different functional groups who specialize in different areas cannot step in until it is their turn and cannot provide the necessary and timely feedback due to lack of visibility into certain stage of product development.

It would be nice to have the opportunity to gauge the changing market conditions and adjust the product development during the execution phase. Agile provides this flexibility through incremental changes of work called *sprints* or *iterations*, where each sprint is a shippable product version. Agile also uses the concept of *scrums*, which is based on feedback and testing of product requirements within short iterations. In this way, the product is developed in incremental baby steps that are scoped and tested well. Agile method uses this short, incremental, and parallel development model to avoid unpredictability. In the agile model, every stage of the product life cycle under planning and execution phases, such as concept, requirements, design, and development is reevaluated and revisited throughout the execution phase on periodic bases. This provides a chance to pause and correct the course to make sure that the product being built is adapted to address the current most requirements, and those requirements are revalidated as well as new ones can be incorporated. In this way, the product development cost and chances for failure is kept low while accelerating the time to market. With the agile model, the development cycles is short, usually limited to couple of weeks, which provides an opportunity to recalibrate software releases and be competitive. So the end product is built right rather than building a product that is not relevant to the market and ends up being a failure despite of perfect development processes being followed.

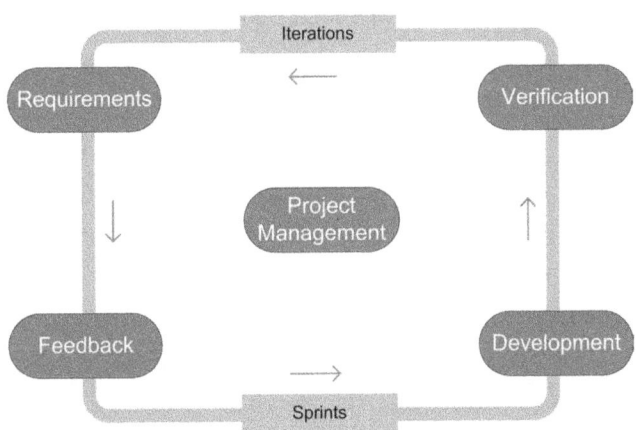

Diagram 4: Agile Methodology

The agile method uses short, incremental, and parallel development model to pause and check periodically and to avoid unpredictability.

There are more details to discuss for both development methodologies, but those are beyond the scope of this book. It is important to note that both the waterfall and the agile methodologies have their pros and cons. The waterfall methodology is considered too stringent and may only suit certain type of products and large-scale companies, probably not suitable for start-ups that need agility and need to adapt as it learns. It is, however, considered a disciplined approach with identifiable stages and milestones that have clear entry and exit criteria and documents everything that is done from quality management and knowledge transfer perspectives. Agile could prove suitable for nonsequential product development in start-up-like companies but rather misfit for large-scale companies. Some companies may choose a hybrid model.

Predesign Activities

Business Case

After the concept commit has taken place, the product manager proceeds with detailed business case analysis. The business case provides the factual reason for why a product should be built at the first place, that is, how it will help increase the business. If you remember the definition of a product from the first chapter, then you would remember that the reason product managers usually build products is to make money. The business case quantifies how much money will be generated. It advocates and supports the reasons why a product should be built. Building a compelling or a lousy business case will determine whether the executive committee commits to allocate required money and resources to build the product or not.

> *A business case provides the clear net profit picture by comparing revenues and costs and justifies why a product should be built.*

Most of the tasks completed at the concept stage feed into preparing a business case, such as total addressable market, realizable market, competitive analysis, price positioning, and revenue forecast. One part of the business case is the margin analysis that has been covered before under the concept proposal. The second part of the business case is the overall *return on investment* (ROI) or *breakeven* analysis. Before the product manager perform the breakeven analysis, one additional item is needed, which is the estimated cost to develop the product. Breakeven is a point in the product life when it has generated enough net revenue to pay back or even out the costs spent on building it. Those overall program level costs are required for the full financial analysis of the project before an execution commit can happen. The overall program costs involve costs in getting the product ready such as the engineering salaries, any regulatory, safety, or compliance testing, certifications, prototype, and other costs.

> *A breakeven point is reached when product has generated enough net revenue equal to the total cost spent on building the product.*

A significant program cost component involved in building a product is *nonrecurring engineering* (NRE) cost. Nonrecurring costs are one-time preproduction engineering related costs to develop a product. Those costs include salaries spent on the resources working on the project, materials, capital equipment, and other such items. The costs for business case generally do not need to include any sales or marketing-related costs but only the costs to complete a finished product. Generally, a program or project manager, who coordinates the new product development effort, referred to as the *new product introduction* (NPI) process, calculates the program costs and helps the product manager finish the business case. It is usually a good idea to overestimate, rather than underestimating the program costs so that the appropriate *funding* can be secured to develop the product.

Once the product manager has the total costs and total revenue forecast in hand, it can prepare the business case and provide a *breakeven analysis*, calculate the *margins* for how profitable the product will be for the company. It is after the breakeven point that a product really starts contributing to the business and starts providing the return on investment. The profit margin, in this context, is the net profit that is obtained by subtracting total costs, including the COGS and productizing costs discussed above, from the total revenue of the product. It is the margin that actually generates cash and is usually tracked by the product managers and general managers as well as CEOs closely as this directly contributes to the company's balance sheet. We will explore this financial relationship more toward the end when discussing the P&L aspects of the product management.

> *The profit margin is the indicator as a percentage of net income generated by subtracting total costs from total revenues.*

EXAMPLE 1

Program Costs

Product	NRE Costs	Proto Alloca-tion	Prototype Costs	Other Costs (Optics)	Head-count	Headcount Cost	Total Cost
10G Fiber ToR	$50,000	20	$60,000	$3,000	3.5	$525,000	$638,000
10G Copper ToR	$25,000	10	$35,000	$0	2.0	$300,000	$360,000
Options	$15,000	10	$10,000	$0	1.0	$150,000	$175,000
						Total	$1,173,000

Business Case

- Total Cost to develop the product: $1.173 million
- Total Revenue during the first 3 years: $126 million
- Breakeven point = Middle of 1QFY1
- Total realizable TAM = $4.2 billion
- Market Share gain = 3%

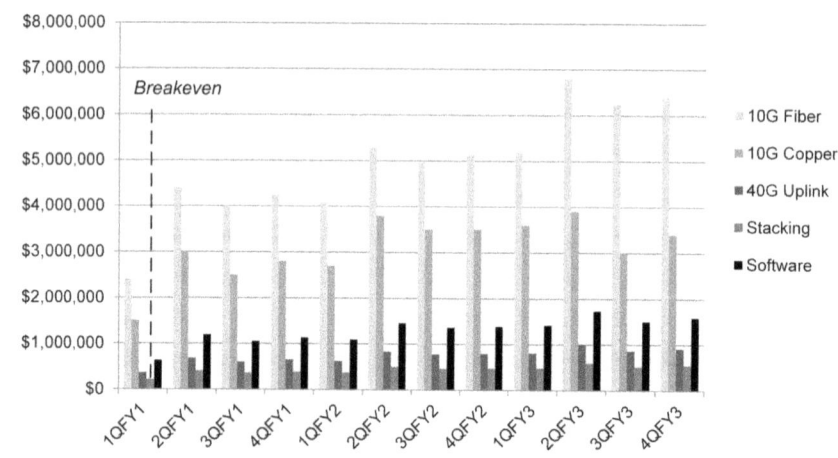

Conclusion

It is totally viable to build this product and asking for the Execution Commit approval because:

- Product reaches the breakeven point quickly—in the first quarter
- Product provides very high return on investment—107X in just three years
- Product provides high margins—62 percent on average

Execution Commit

The execution commit (EC) milestone marks the beginning of the execution phase. There are several things that are usually needed to be completed before an execution commit or *stage 1 exit* can take place, such as:

- The product requirements document has been completed, reviewed, and approved by all stakeholders.
- The business case has been completed and reviewed.
- Required resources have been identified for the project.
- Required funding has been secured for the project from the R&D budget.

By the execution commit, the product manager brings everyone on the table to make commitments for the time and resources being requested to build and launch the product. As result of initial scoping and feasibility analysis after the concept commit, the stakeholders committing to the project have perhaps done their own homework by the time whether they can support the effort or not and to what extent. Negotiations between the product manager and the core team are norm rather than exception before or at the execution commit. Below, we will explore some of the key topics from the list of tasks above.

> *The execution commit (EC) is the stage of a product life cycle process at which a new product development project is committed by all functional groups involved and required resources, funding and time to deliver the product is allocated.*

New Product Introduction Team

Once the execution commit has been approved, resources allocated by different functional groups shape up a project specific functional team, usually referred to as the *new product introduction* (NPI) team or the *core team*. The core team exists only for the duration of the project, and it is the core team's charter to execute upon the requirements that the product manager has asked for and to deliver a finished product according to it. Generally, the person in charge who is designated for driving and coordinating the core team activities on daily and weekly basis is called a core or *NPI program manager*. The program manager works closely with the product manager to get things done.

The working relationship between the product manager and the core program manager should be very healthy. It is the program manager's primary responsibility to support the product manager in getting the product out of the door on time and to remove any obstacles. Therefore, the program manager's role is of a middleman and a mediator. The issues can arise when the program manager considers himself on the side of the functional teams and starts questioning the requirements or starts defending and justifying any delays or misses by the core team, which is totally counterproductive and unfortunately happens too often. The program managers need to understand that by the time requirements come to the core team, company executives and key stakeholders have already approved and funded the idea. Therefore, the core team's job is not to question the rationale of building a product, but rather executing upon it to deliver it on time. To avoid this problem, many companies maintain an independent *project management organization* (PMO) that does not report into a functional group. This helps in maintaining a neutral and balanced posture of the program management with tracking and control in managing resources, costs, and schedule.

Design and Development

There are several different activities involved in designing and developing a technology product. For example, in case of a networking product that could be both hardware and software based, there are some key design and development activities that must be completed regardless of what type of methodology (waterfall or agile) is used. Below we will explore some of those activities.

Mechanical Design and Development

In case of most hi-technology products, there is some sort of mechanical design involved. For example, in case of networking products such as a switch or a router, the mechanical design involves designing the physical chassis form factor, sheet metal related work, slots, modules, latches, handles, cable routing, airflow paths and power distribution, etc. Additionally, the mechanical design of the product needs to consider safety and compliance standards for fire suppression, thermal, vibration resistance, airflow, acoustics, failover, and fault tolerance. The mechanical design may also need to comply with certain industry-based standards. For example, in the networking industry, *network equipment building system* (NEBS) is a well-known standard that dictates safety and environmental guidelines for the IT equipment manufacturers. NEBS have multiple levels of compliance ranging from level-1 with minimum compliance, to level-3 with highest compliance.

It is important that while considering the mechanical requirements, the product manager carefully surveys or understands the environment in which the product will be installed and also how it will be handled. For example, the physical chassis size and form factor in terms of chassis height, width, and depth does matter in the networking industry. This is important for the target use cases. If it is a Top-of-Rack (ToR) switch, which needs to fit into a data center standard server rack, it has to consume minimum space so typically it is one or two *rack units* in height. A rack unit is 1.75 inches. If it is a core switch, it can be a large chassis several rack units high to accommodate more density for aggregating multiple ToR switches. Similarly, a service provides router could be pretty large size. While dictating the form factor requirements, other factors such as chassis weight, depth, and layout should also be considered. For example, if the chassis is mounted on top of other equipment or if it is mounted in a two post rack, then the weight will matter. Also, the chassis depth should be considered for fitting into the standard racks and cabinets.

Another important consideration in case of a chassis is whether it will be *fixed* or *modular* form factor design. And if it will be modular, then what will be the design and orientation of the *interface modules*? Which components will be populated and removed from the front or back of the chassis? Given that the product design is modular, other factors will need to be decided such as how many empty slots it will have for housing different component options such as the interface modules or power supplies. Whether those modules will be full slot or half slot wide, and whether they will be horizontally or vertically oriented. The layout and the orientation of the slots will affect how easy it will be adding or removing the components when cabling is in the way and also how the cables will be routed in dense deployments.

A fixed form factor design does not allow pluggable or swappable components; therefore, it is less flexible but lower cost in manufacturing. A modular form factor allows flexibility by adding or removing the components on demand and therefore customizing the product according to individual customer needs. Obviously, this design is more expensive to design and manufacture, but it could also be priced higher. If a product is designed with wrong dimensions or form factor for the target installation environment or market, it will make the product unusable. When it comes to the form factor, it is true that one size does not fit all. Usually, a mechanical engineer on the core team or the contract manufacturer handles those design activities. Detailed discussion of mechanical design process is beyond the scope of this book. From the product manager's perspective, what matters is that the product fits the requirements in terms of form factor, dimensions, layout, modularity, and weight.

Industrial Design and Branding

It may not seem so but it is of significant value to carefully design the cosmetic look and feel of the product. Things such as the paint color, texture, panel design, any bezels, brand logo, product name, lighting, any covers or doors, and other cosmetic items all add up to the perception of the product. Collectively, this is referred to as the *industrial design*. The industrial design does affect the product's brand. A product's *brand* is a unique item such as a product name, a symbol or a logo, or a design that distinguished one vendor's products from another vendor's products of the same type. The product branding refers to applying the above mentioned attributes to a product. The brand takes time and effort to develop and becomes a primary source of customer perception for judging the quality and experience of a product before even using it. Successful brands precede the products, and products sell because of their brand. Sometimes, a company is identified because of its product brand. In this way, a brand becomes a company's valuable asset. It needs to be protected and promoted.

> *A product's brand is a unique item such as a product name, a symbol or a logo, or a design that distinguished one vendor's products from another vendor's products of the same type.*

While setting the requirements for the industrial design, the product manager needs to keep the corporate branding guidelines in mind, usually dictated by the corporate marketing team. This applies to using the company logo, selecting the paint color, any registered trademarks, and other such items. However, product branding is not limited to the industrial design only. Product names used in any *software command line interface* (CLI) and logos used in any *graphical user interface* (GUI) are also part of the product branding. Additionally, designing the product packaging and any labels that go on the product and its package also need to consider the branding rules.

EXAMPLE 2

Mechanical and Industrial Design for the ToR Switch

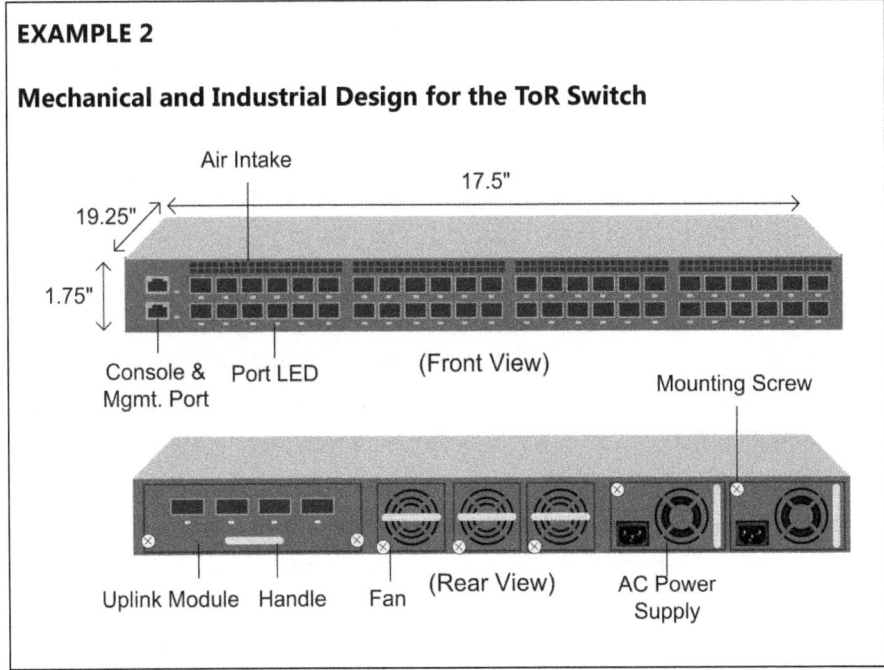

Power and Thermal Design and Development

Some of the most important aspects of designing a networking product include its power and cooling system design. How much power budget will be needed for the product at its peak operation? How will the power be distributed and consumed? Power options such as AC and DC power, power redundancy for the power supply, or power source failure protection are all important questions that the design needs to solve. The product should also be designed with enough power budgets for the future expansion and addition of new components. In addition, the electronic and moving components used in the product generate significant amount of heat under operation. On top of that, part of the power consumed is also dissipated as heat. The total heat generated raises the component level and ambient temperatures around the product that must be cooled down to keep the temperatures below the maximum operating limits of the components. The product design therefore needs to consider how the product will be cooled.

Part of the cooling is how the airflow will be directed over the components generating heat. For example, in data centers, front to rear airflow is mostly a requirement for the network switches because data centers house large number of equipment that generates lots of heat. For this reason, data centers maintain the *hot* and *cold isles* physically segregated with cool air from the chillers flowing in and the hot air collected from the equipment and exhausted. The product design needs to have vents on the front of the chassis and fans on the back of the chassis that pull the cool air from the front into the chassis, which cools down the hot components while it collects the heat and exits from the rear. In addition to the airflow, other cooling efficiency aspects such as where to place the heat sinks and how to control the variations in the fan speeds are some of the other factors to be considered in the thermal design.

Every product has a minimum and maximum operating temperature threshold that the product is tested to operate in. Part of the thermal design is to place thermal sensors at the strategic locations within the product such as on or next to the critical electronic components and heat sources. The sensors monitor and collect the temperature information and feed into the central software logic making decisions about the product thermal operations. The software may as a result increase or decrease the fan speed or shut down an overheat section of the product and generate an alert. This protects the expensive product components from failure. Such intelligence in the product also helps in case the ambient temperature rises where the product is installed, such as nearby fire at the facility. Thinking through such cases and defining the thermal requirements helps in high customer experience. Power and cooling efficiency is becoming increasingly critical for the selection of *green* products because of more environmental conciseness and because of its impact on the *total cost of ownership* (TCO) for the customers.

Architecture Design

Once the project execution commit has taken place, concrete product design effort starts. Generally, by the time, work on early architecture of the product should already be in progress. The product architecture design is done under general guidelines and requirements with participation of the hardware and the software teams. Regardless of the type of product, during this stage, the product manager will require from the engineering teams to produce detailed design documents called *functional specifications*. As discussed earlier, the engineering teams respond to the product requirements through a *product design document* (PDD) or set of deviations, which provides the response to the requirements in terms of what can or cannot be accomplished and what changes will be required to the product design that is different from what the product manager has asked for.

The product manager reviews the design document or deviations and makes sure that what engineering and other teams are going to deliver is acceptable by the customers. Any disconnects need to be removed and any loopholes need to be closed during this process. Therefore, the design document sets the basis of a high-level product design. Alternatively, those responses or changes can be captured within the product requirements document itself as deviations, as discussed earlier during the planning phase. Either way, this should be done and agreed upon prior to the execution commit, or it will result into a *scope creep*. The idea is that everyone is in complete agreement before a commit happens. If significant changes are made to the requirements after the execution commit, a *recommit* may be required by the effected stakeholders. What happens from thereon during the execution phase is discussed below in terms of hardware and software design since our focus in this book has been the networking products.

Hardware Design and Development

Hardware design generally starts with a high-level architecture design in which engineering team and the product manager discuss the important capabilities that must be built into the design per the requirements. The discussion at this stage is not around *what* is needed but *how* it is going to be implemented as there could be more than one way to implement a requirement. Usually, several whiteboard sessions and discussions take place between the product manager and the engineering team until a design is fully locked. The product manager's role is not to really dictate deep engineering details but rather act as a consultant and a reviewer to validate what they suggest. If the product is a network switch, as in our examples, the design starts with a discussion such as whether to build it around a merchant or custom *Application Specific Integrated Circuit* (ASIC), also referred to as the *silicon* or the *chip*. The product price positioning and cost targets have an influence on this choice as well. For example, building a new custom ASIC is very expensive undertaking and takes long time to design, build, and test an ASIC, which involves millions of dollars of investments and lots of effort. A slight error in design or fabrication process could result into irreversible damage. Therefore, the custom silicon suits for purpose built, high volume, premium products where there is no other way.

On the other hand, the merchant silicon is usually designed with economies of scale in mind and has an evolutionary roadmap and support mechanism in place provided by the silicon vendor. Merchant silicon is suitable for building products that require faster time to market and require competitive price-positioning strategies. Merchant silicon is getting very advanced in terms of maturity, features, and reliability where it is now being competitive to the custom silicon. However, with the merchant silicon technology there are strong dependencies on the timely delivery of features and functionality by the silicon vendor and its alignment with the market trends. Any changes, or delays, from the vendor may impact overall product design or time to market significantly. Unfortunately, this happens too often. Therefore, one of the many challenges that the product manager has to deal with is to push the silicon vendor's roadmap in the desired direction. Often, the same silicon vendor may be supplying the same silicon to multiple companies who compete with each other, so having a dual-supplier strategy may prove useful in many ways for the product manager.

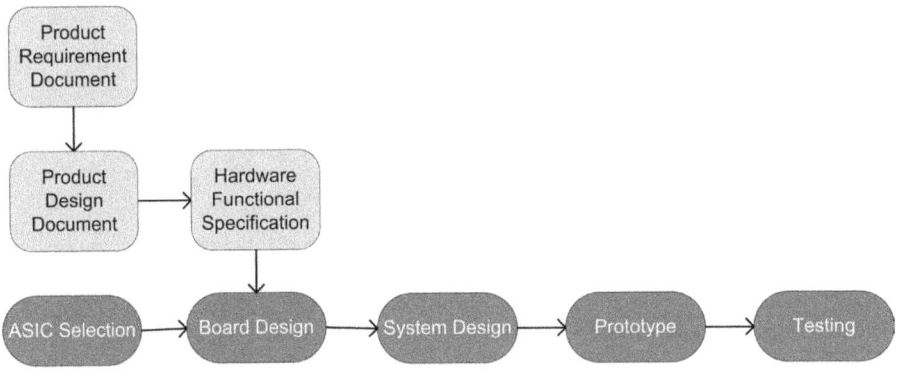

Diagram 5: High Level Hardware Development Process

Although the silicon acts as the foundation on which a network product will be built, such as its speeds and feeds and features, but it does not restrict the innovation. Even if a competitor chooses the exact same silicon, how the product's form factor, hardware, and software are designed and what features are brought to the surface will shape up very different product outcomes. Therefore, within the scope of what the merchant silicon delivers, there is enough room for driving the innovation and differentiation. After silicon has been selected, engineering team follows with the detailed design. Generally, the first stage referred to as the *board design* follows. The board design involves designing the electronic circuit board around the silicon. This is documented through a *hardware functional specification* that is reviewed and approved by the product manager among other stakeholders. In case of *on-demand manufacturing* (ODM), the ODM vendor could also handle the initial hardware design instead of an in-house engineering team. In which case, the engineering team usually acts as the intermediary between the ODM vendor and the product manager. We will discuss ODM model later in more detail.

EXAMPLE 3

Hardware Board Design for ToR Switch

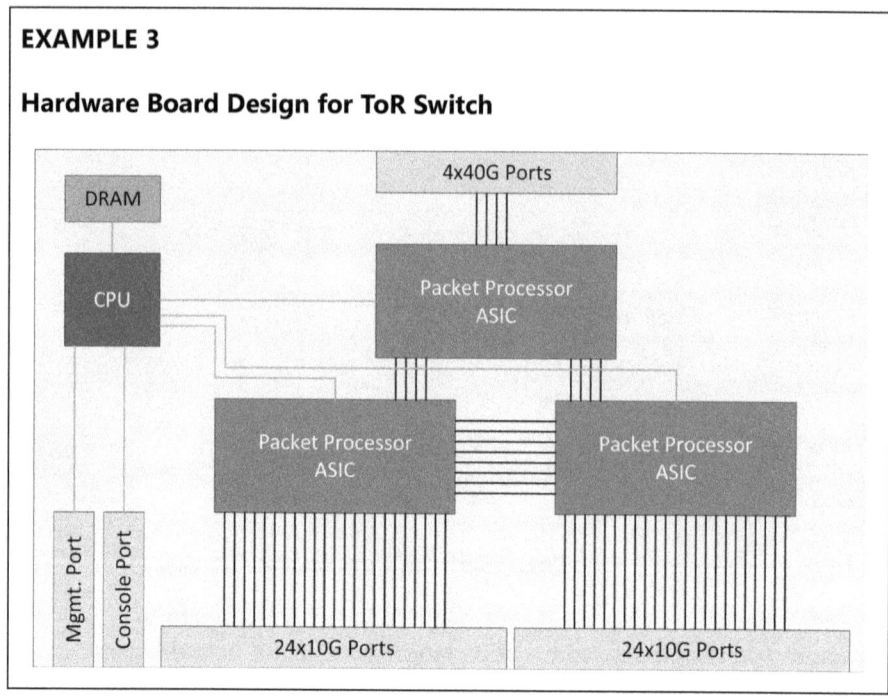

The engineering team also works with the other teams to evaluate the thermal and electromagnetic interface (EMI) impacts of the design. Once the functional specification has been approved and design finalized, the development activities can be started. The very first product that is actually built is referred to a *prototype* or sometimes referred to as a *prototype level-zero* (P0). A prototype is an early stage "sample" product built to test the product functionality and carry forward the on-top product development. The prototypes are expensive compared to the actual production grade product because they are very few in numbers and a manufacturing assembly line has not yet been set up, rather those prototypes are prepared by the engineering team. The first prototypes play the most critical role in the development process. This is the first chance for the engineering and other teams to play with the real product and uncover any issues and fix those issues. It is extremely important that any hardware defects are found at this stage; otherwise, more respins will be required with more changes in the hardware and costing more *research and development* (R&D) dollars to the company.

Based on the complexity of the design, there could be a planned respin of the prototype. Once the P0 prototypes have been tested and any necessary changes have been made to the design, the next step is to build the planned *prototype level-1* (P1). The second prototypes are still limited in quantity but better in stability and used for verifying the changes and fixes made to the first prototypes. Once the product design and functionality has been verified and no more changes are expected, the next stage is to build the beta units for winder distribution to internal teams and to external customers for testing. At this point, the hardware development should be pretty much complete if everything goes well. The beta units are built on the manufacturing assembly line. We will explore more of this in the following sections.

Software Design and Development

Software design also starts with a high-level design first. However, the software detailed design generally cannot progress fully until high-level hardware design has been completed, but it really varies from product to product. The hardware and the software teams usually work together to take the product design forward. The software team locks the detailed design by writing a *software functional specification* document that must be reviewed and approved by the product manager to make sure that what is going to be delivered is according to the requirements. Unless the product is a software-only product, the software team needs real hardware in order to start the software *integration* process on it. Therefore, software work cannot progress until the first hardware prototype (P0) has been built. The software integration involves low-level tasks such as writing *embedded software,* such a *device drivers* to "bring up" the new hardware. In case of merchant silicon use, low-level software routines are usually provided by the silicon vendor in the form of a *software development kit* (SDK) that includes many software *libraries* and *application programmable interfaces* (API). This makes it easier and saves time for the software teams to focus on the purpose-driven software development only. However, this also bounds them to the capabilities supported by the software development kit.

The first prototypes are used by the software team or the ODM vendor to complete the software integration and run the diagnostic testing. The software team then starts *coding* the high level software such as required product features and functionality according to the requirements. This continues on the first and the second prototype builds. When the software development is complete, a milestone referred to as the *code complete*, the *unit testing* starts by the software developers to test the low-level integrity of the code. Unit testing is also referred to as the *white box testing*. As we will explore later, as the hardware and the software development reaches completion, next stable revision of the product called the *beta* version if built. The beta product is built more in number compared to the prototype build for distribution to broader engineering, testing, and compliance teams. The software developers fix the known defects or "bugs" found in their code and then hand it over to the focused test teams for the *quality assurance* (QA). We will explore the quality assurance process in later chapters under sustaining discussion.

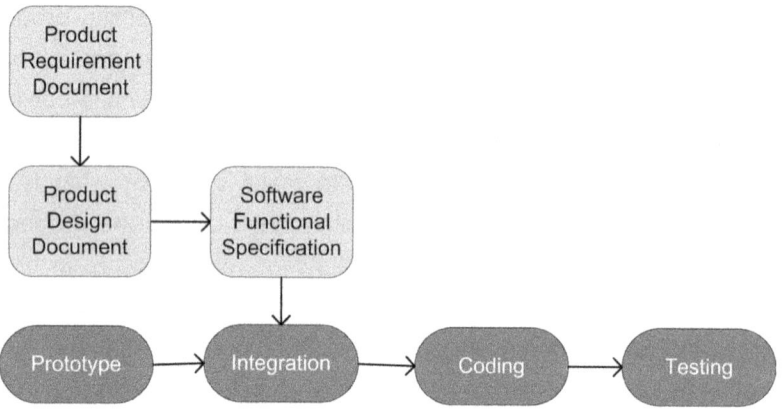

Diagram 6: High Level Software Development Process

The quality assurance or functional test team performs the *black box* testing to validate the product functionality and behavior as expected by the customers. The product may also be put through the *system testing* in a more complex environment that simulates closer to the real deployment environment to uncover and fix hard to find defects. Additionally, the product may go through a *regression testing* cycle to make sure that any software features previously supported still work for the backward compatibility and no "collateral damage" has been caused by the new software development. Once the required level of product stability has been achieved, the new product is ready for the real-time customer testing or a *beta trial*. If the product is purely a software product without requiring any hardware integration, the sequence is similar. The high-level design is followed by a detailed functional design, followed by the coding and then testing.

Prototype Build

In case of a hi-technology product, once the hardware design has been finalized, a handful of sample product units can be fabricated for the hardware verification, software integration and further development activities. Those samples are referred to as the product prototypes as discussed earlier. As discussed earlier, a prototype is an early sample before the manufacturing can start. It is used to test the product functionality and carry forward the product development. The very first product that is actually built is referred to a as a *prototype level-zero (P0)*. The prototypes are expensive compared to the actual production grade product because they are very few in numbers and a manufacturing assembly line has not yet been set up, rather those prototypes are prepared by the engineering team.

> A prototype is an early stage "sample" product built to test the product functionality and carry forward the on-top product development.

The first prototypes are usually built for engineering use only so that if there are any serious hardware or software issues, now would be the time to find and fix those. It is much more expensive, damaging, and limiting if a hardware issue leaks out to volume production stage and widely spreads in the customer install base. Even if the defect could be fixed, hardware respin will be required costing valuable time and money to the customers. The affected components will need to be collected and sent back to the factory for rework, which will result in interruption in customer operations and bad reputation of the product. It is also possible that the problem may not be fixable at all at that point making it worse. In which case, resulting in a crippled product operation or replacing the product. For this reason, there is usually a planned respin of the prototypes, the *prototype level-1* (P1). The second prototypes are still limited in quantity and used for verifying the changes and fixes made to the first prototypes. Depending on the need, risks, and complexity of the product, there could be even another respin of the prototypes planned, the *prototype level-2* (P2), but it is rare.

Emissions Testing

Another use of the prototype build in case of a network product is to test the radio frequency (RF) energy emissions through the product. This is referred to as the *radiated emission scan* that is usually tested in a controlled chamber environment. The goal is to find any *electromagnetic interference* (EMI) issues and make the necessary changes to the board design to correct it before it is too late, although the detailed EMI testing is performed later in the stage as discussed in the following sections. Detailed discussion of these deep engineering topics is beyond the scope of this book.

Manufacturing, Testing, and Conformance

Beta Build and Customer Trial

As the product development and the basic internal testing have been mostly completed, the next stage is to build the beta units for wider distribution to internal teams and to external customers for testing. The beta units are generally built on the manufacturing assembly line, so they are sort of first preproduction build as well. In case of software-only product, the beta build is easier and usually only happens once since there are no prototypes built and no respins required. Typically, the beta units are internally distributed among the hardware and software engineering, compliance testing, platform and regression testing, and other teams who need to finish their part or validate the product timely. However, the beta units can also be distributed externally to selective set of customers for their trial.

The objectives of beta trial are twofold. First, to test the product in a close to target deployment environment. This helps collect early quality feedback in terms of customer experience and uncover any technical issues that could be fixed before wider customer base is exposed to them. This is the engineering value out of the beta test process. Second, to let the potential customers familiarize with the product and have them experience it, which can result in potential sales and word-of-mouth marketing. This is the marketing value out of the beta test process. If either of the value cannot be obtained, then the beta testing may not make much sense.

> *A beta product is the first manufactured product used for internal and external testing, for regulatory compliances and certifications.*

The beta trial is usually done with narrow spectrum of functionality in mind and under close guidance from the engineering team since the product is technically still under development and not fully complete. Any caveats are clearly and transparently communicated to the beta customers. Usually, the engineering and the internal test teams come up with a proposed *beta test plan,* which is provided to the customers willing to test the product as a guideline. Another important aspect is to choose the right version of the software that is stable enough for the beta testing. Providing an unstable version to beta sites may ruin the very first customer impression and customers may run into irrelevant issues that block them from testing the features they are interested in. Since the software is usually still going through testing as well as *bug fixing* process internally, it may not be fully stable. Therefore, a combination of stable-enough code at the fundamental level and in the feature areas targeted for the beta testing, combined with a guideline test plan, may prevent customers from running into the mine field where they do not need to go. It is not unusual to provide software updates to the customers on periodic basis during the beta testing.

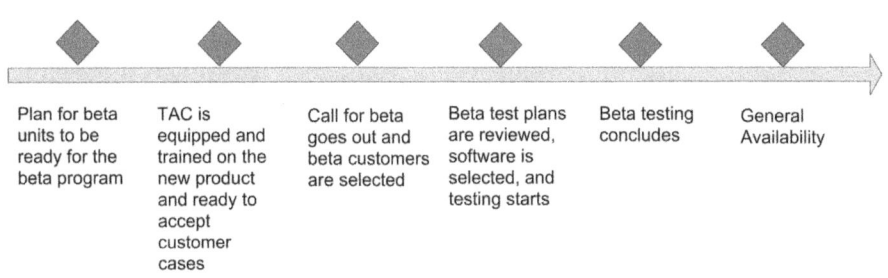

Diagram 7: Typical Beta Program Steps

It is important to select the beta candidate customers carefully. To do this, usually the product manager issues a *call-for-beta* announcement to the sales team and partners, asking for candidate customer list who would like to try the new product and "play with the new toy." Sales team finds and lines up those customers who are interested in trying the new product. Also important is to line up any resources to support the beta testing process from end to end, such as the *technical assistance center* (TAC) to provide adequate technical support and to open the *trouble tickets* on customer's behalf if customers find any bugs (which is highly desirable). There should also be adequate engineering resources assigned to troubleshoot and fix the reported issues in timely manner. Before kicking off the beta test program, the TAC team should have the necessary test equipment and the new product installed in their labs as well as trained on it so that they can quickly *reproduce* the reported issues in-house for engineering to troubleshoot and fix them.

Usually, limited set of preliminary product documentation is also made available for the beta customers to get started. Another milestone that may or may not be executed upon in-line with the beta trial is the opening the product for sales quote referred as to the *open-quote,* which consists of providing the preliminary product pricing to the potential customers and start issuing the sales quotes to customers who show interest in buying the product. A *sales quote* is a price estimate per requested quantity of a product. The open-quote is a milestone related to product launch process and therefore will be discussed under the appropriate chapter in detail later on.

Regulatory Compliance and Product Certification

The beta product units are also used by the safety and compliance team's testing for the product safety, environmental, emissions, and immunity-related issues against the standards and obtaining the required certifications necessary per different local and international agency regulations. Typically, the three key areas that are tested by the compliance team include:

- Safety testing—involves testing product safety related issues such as testing for high/low temperature, electric shock, electric short, wiring, drop test, etc.

- Electromagnetic interference (EMI) testing—involves testing for full RF emissions, electrostatic discharge, conducted and non-conducted testing, etc.
- Environmental testing—involves testing for high/low temperature, humidity, shock and vibration, etc.

Some of these certifications must be marked on the product and package labeling to indicate that the product is fully complaint with the safety and environmental standards.

Trade and Export Compliance

Since the beta units are the first type of units that could possibly be exported to other countries for customer trials, they also need to be cleared by the local and the international governments for obtaining the import and export certifications. Generally, a product cannot be exported, imported, and sold into other countries without the right licenses; a process also referred to as the *homologation*. The global trade service (GTS) team provides the required information to the commerce department for obtaining the product classification, also referred to as the Commodity Classification Automated Tracking System (CCATS). This process generally starts way before the beta milestone and does not actually consumes any beta units, but it should be completed by the beta milestone. The process is especially important if the product involves any security or encryption features.

Product Quality Assurance

A company's *quality assurance* (QA) process and its caliber have a direct impact on the product quality, its perception, and reputation; all of which affect both business and customer experience. The primary job of the quality assurance or testing teams is to uncover as many defects as possible before a product or a software release goes out in customer hands. This way, the dirty laundry stays home, and also it is more convenient and less expensive to fix those issues in time, as well as to prepare to message them to customers if they cannot be fixed soon. If a customer finds a defect during real-time operation of the product, it is not pleasant. It hurts the product and company reputation. When the beta units are built, they are also allocated to different test teams for verifying the product features and functionality and to uncover any issues before customers do.

The test teams design the test beds to simulate the customer environments and write test cases to simulate different situations and scenarios under which the product will likely be deployed. For example, the functional test team verifies the product features and functionality and expected behavior. The product is tested positively and negatively for the exceptions. The system test team puts the product into broader solution to simulate complex real-world environment and verifies any interoperability issues as well as tests the product under stress and closer to maximum capabilities. The regression test team runs the automated test scripts or manually verifies the product functionally for any backward compatibility and to find any collateral damage in software. The quality assurance process is a vast subject in it but beyond the scope of this book.

Pilot Build

Once the hardware development, prototyping, software integration, beta trail, and most of the testing and compliance have been completed, the next step is to start building the *pilot* product units. A pilot product is almost a production grade product that is basically a candidate final for the shipment but still limited in quantity. The pilot product units go through the preliminary manufacturing and packaging process to assess how the finished product will look like and to assure the full *customer experience*. One of the most important reasons for building the pilot units is to conduct what is called the *first article inspection* (FAI). The first article inspection is a process in which the components are closely inspected and installed for catching any last-minute anomalies when they are first received from the manufacturing facility. This ensures that everything from physical look and feel to the installation and packaging is as expected before the volume manufacturing and shipment to customers starts. Product packaging, labeling, and documentation are also inspected. The process ensures the full customer experience. Pilot units are also used for sales and technical support teams training.

> A pilot product is almost a production grade product that is a candidate final for the shipment and used for last-minute complete checks.

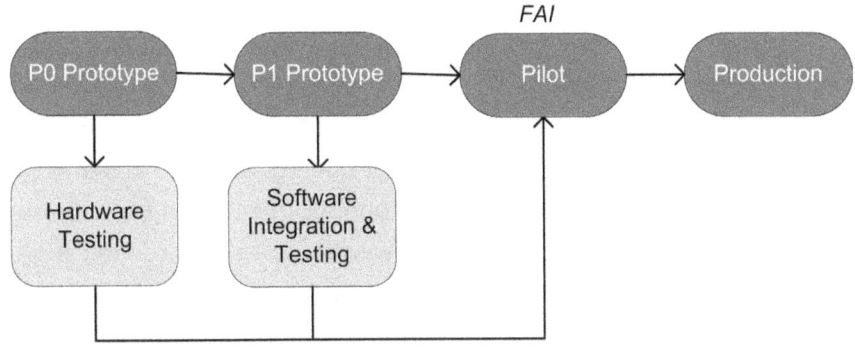

Diagram 8: Hardware and Software build phases

Services Readiness

One of the uses of the pilot products is to equip the technical support or the *technical assistance center* (TAC) team that is part of the services organization so that they can set up their "test beds" and can be trained on the new product deep down. After a product has been launched and if a customer finds a defect in the product, its first contact is generally the technical support team. The customer usually calls the technical support or files the issue through an online reporting tool based on which initial investigation takes place between a technical support expert and the customer. Relative information is gathered that will be useful to debug the issue later on to find the root cause issue, and it is relayed to the engineering team. The engineering investigates the issue and provides a fix back to the customer through the technical support.

However, before an issue can be fixed, it must be recreated or reproduced, which means that the necessary customer environmental variables are simulated as close as possible to understand how and why the problem occurs. This why the technical support teams need to have simulated customer environments setup ahead of time, and they use the pilot products for this purpose.

Demand Forecast

Before the product starts manufacturing or selling in volume, the product manager needs to provide the marketing forecast to indicate the product demand. This forecast is consumed by the operations and manufacturing teams or the contract manufacturer to determine how many units of product need to be manufactured in volume every quarter. This initial forecast may or may not be based on any *sales guidance* and may not be very accurate, but that is usually expected. If the product open quote has happened, and if the sales team might have issued any quotes, there could be some sales guidance for preparing the forecast. However, the sales guidance at early stage, especially when the product has not launched yet, may be nonexistent; and therefore, most of the initial forecast comes from the product manager himself. The product manager can use other reference information such as actual shipment volumes of any existing or past comparable products or any intelligence regarding similar competitive products market share. If the new product is a replacement of an older product, then it is relatively easier to forecast as the market share and the actual shipment volume are known.

The initial forecast is generally transferred from the product requirements document into a forecasting tool that a company maintains for this purpose. From thereon, the forecast is maintained in the tool and is a recurring sustaining activity, as we will explore later. Before the product is shipped the very first time, referred to as the *first customer shipment* (FCS) or the *general availability* (GA), a forecast for first three years is usually needed for proper demand planning. As the product starts selling and a trend starts taking shape, the forecast accuracy can be greatly improved. The product manager also looks at the quarterly *sales targets*, *sales pipeline*, and *bookings* to use as inputs into the forecast. We will explore these concepts more under the sustaining phase chapter.

Based on the initial forecast, the operations team starts planning the supply chain with the suppliers and starts determining the component level costs. The operations team, also being responsible for the shipment of the product, looks at the forecast to determine how many units need to be manufactured to fulfill the projected customer demand in the upcoming quarters and weeks. Once the right number of product units has been manufactured, they are shipped to the warehouses around the world, from where the regional orders are fulfilled. The operations team also maintains a *safety stock* for the emergency purposes in situations when a large unexpected order comes in and depletes the inventory.

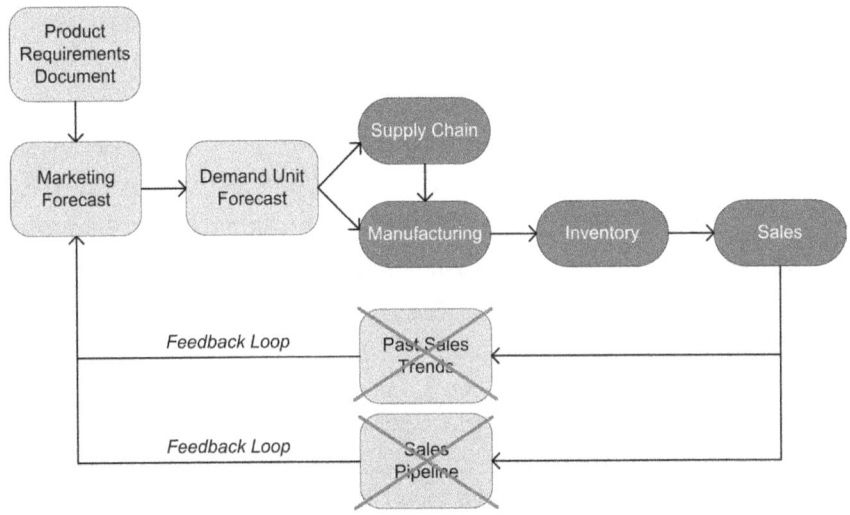

Diagram 9: Initial Product Forecast Process

Inventory management is an important activity that needs to be managed on an ongoing basis. For healthy financial performance of a company, there should be enough product units manufactured in a given quarter that ideally are fully consumed by the last day of the quarter. However, obviously it cannot be predicted with that much accuracy. In many cases, especially when a product is new and historic trends have not been yet established, an *over-inventory* situation can occur. Over-inventory is not a good situation because in the company books it is considered an expense. Storing the product units in warehouses costs the company, especially in the case of third-party manufacturing. Therefore, the operations team periodically monitors the inventory levels with the product manager and adjusts the demand forecast accordingly.

The forecast for a successful product is expected to go up every year, if not every quarter, as do the sales targets. Increase in the product demand indicates success of the product and good health of the business. The product manager is usually required to present the forecast to the executives on monthly or quarterly basis and is answerable for any downward shifts in the forecast. This could be perceived as some sort of issues with the pricing or functionality of the product, sales execution, marketing, channel partners, or other areas that will need to be investigated and fixed.

In case of the software products, the process is very similar, except that the items being forecasted are shippable software entities such as licenses or downloads. It should, therefore, be clear that forecasting is a very important task. We will discuss more on forecasting subject in the later chapters.

EXAMPLE 4

ToR Switch and Options Demand Forecast

Fiscal Quarters	Projected Unit Volume				
	10G Fiber	10G Copper	40G Uplink	Stacking	Software
1QFY1	300	150	225	270	315
2QFY1	550	300	425	510	595
3QFY1	500	250	375	450	525
4QFY1	530	280	405	486	567
1QFY2	510	270	390	468	546
2QFY2	660	380	520	624	728
3QFY2	620	350	485	582	679
4QFY2	640	350	495	594	693
1QFY3	650	360	505	606	707
2QFY3	850	390	620	744	868
3QFY3	780	300	540	648	756
4QFY3	800	340	570	684	798

Product Structuring and Supply Chain Management

Before the product goes to volume production, it is critical that all source suppliers who will provide the necessary components used in the manufacturing of the product have been identified and required quantities of those components have been secured per the demand forecast. Also, that the costs for those components have been negotiated and finalized. This is done through the *commodity* or *supply chain* team that is usually part of the operations team. Generally, the process begins by the product manager submitting the new marketing part numbers for the product into a *product structuring tool*. Different companies use different processes and tools. One such commonly used tool is agile.

Every product has a *marketing part number* assigned to it, which is meant for the external use such as ordering of a product and is published in the company's price list, as well as linked to the *channel partner* ordering systems. In addition, there is also an internal *manufacturing part number*, which is used internally and never visible to the outside world. Every manufacturing part number has one or more *bill of material* (BoM) attached to it. Every component used in the product or its packaging is basically a bill of material, and all supply chain negotiations happen at the bill-of-material level. Example of a bill of material could be as small as a capacitor, a merchant silicon kit, a package, a label, a software license, or as big as a preassembled third-party product. Therefore, from operations perspective, a product is usually a tree of bill of materials rooted at the manufacturing part number level. The product structure could be a nested tree as well. The top-level "parent" bill of material could consist of several "children" bill of materials, each of which could consist of other "grandchildren" bill of materials.

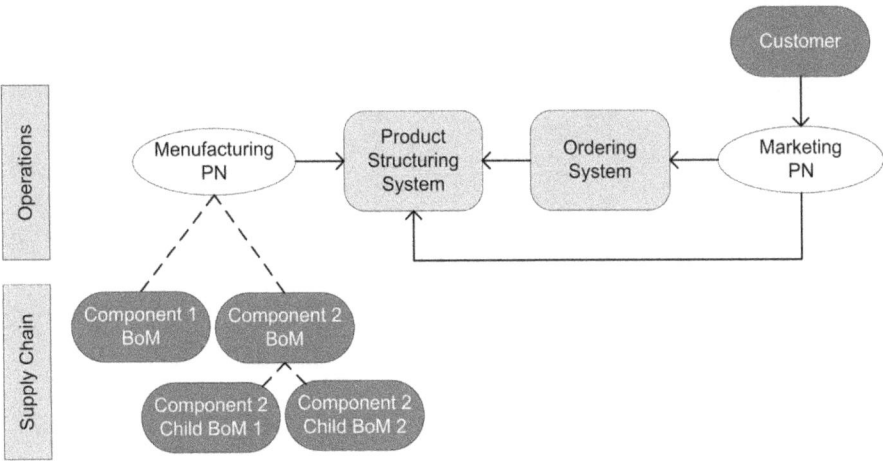

Diagram 10: Product Structuring Process

It is the product manager's job to provide the data to the appropriate teams for building the initial product structure. Usually, the team responsible for the product structuring is called a *product data management* (PDM) team. The product manager starts the process by filling the data in some sort of online or offline form per the company's product lifecycle process. This step is called a *bill-of-material request*. The form requires data such as the marketing orderable part number, product description, product family, internal project name, serialization requirement (having a serial number or not), Universal Product Code (UPC) requirement, etc. The product data management team allocates a manufacturing part number and assigns a *process engineer* who structures the product with bill of materials in a structuring database tool. Once structured, concrete links among multiple components have been established as an overall single entity—the product.

Any time a change is made to the product structure in the database, it is called an *engineering change order* (ECO) process. Since the product structure is an important entity, any changes are made under a controlled change management process by initiating a changer order request that is tracked and approved by the key stakeholders including the product manager. Once the changer order has been approved and the required changes to the product structure have been made, the changer order is complete. From thereon, every time a change is requited, another changer order must be initiated, processed, approved, and recorded.

> An *engineering change order (ECO) is a controlled change management process to make any changes to the product structure such as components or work flow.*

EXAMPLE 5

Bill-of-Material Request

		Initial BoM Request			
Number: ECO-123	**Business Model:** Z-Series	**Requestor:** Nadeem Zahid	**Requestor's Job Role:** Product Manager	**Request Date:** 15/08/2013	**Target Complete Date:** 30/08/2013
Marketing PN	**Manufacturing PN**	**Product SKU**	**Long Description**	**Product Line**	**Serialized**	**UPC**
101	300-1111	TOR-10G-48F	48-port 10GbE SFP+ ToR Switch	Z-Series	Yes	
102	300-1112	TOR-10G-48C	48-port 10GbE RJ45 ToR Switch	Z-Series	Yes	
103	300-1113	TOR-UL-40G-4	4-port 40GbE Uplink Module	Z-Series	Yes	
104	300-1114	TOR-STACK-480	480Gbps Stacking Module	Z-Series	Yes	
105	300-1115	TOR-L3-LIC	Layer 3 Feature License	Z-Series	Yes	

Once the product structure is complete and the demand forecast is in place, operations and commodity teams take it and derive the demand for every bill of material appropriately. For example, if the product manager forecast for a product is 100 units in a certain quarter but it uses a certain component in quantity of 4 for each unit of product during manufacturing, then the commodity team will communicate a demand of 400 of those components to the supplier plus few extra for replacing any bad components. The commodity team negotiates the costs based on the component level forecast, which becomes a contributing factor to the overall raw-COGS for that product. Negotiating the costs for high-cost materials periodically with existing suppliers, finding alternate sources of supply, and shopping for more efficient components and material is the primary job of the commodity team, and it affects the product margins positively or negatively.

Stock Keeping Units

The time a product is being structured is also the time to properly design product family tree and think about how different product *stock keeping units* (SKU) will be created and managed over time. A SKU is simply a sellable iteration of the product under its own orderable part number. A product could have multiple SKUs for identifying multiple versions or variations of it. The SKUs could be based on different designs or just some variations to the same product to suit different applications. Every SKU is identified usually by a unique part number and description and has its own price. A SKU on the back-end could reuse the very same or different combination of BoM and quantity.

Only certain product SKUs of the same product will be usually successful in the market and not all of them. Therefore, the product manager should give some thought before creating lots of SKUs. Every SKU needs to be forecasted, stocked, and maintained as standalone entity will later on require inventory management. It is a common pitfall for the product managers to define too many product SKUs initially and later on deal with ongoing SKU and inventory management issues during the product sustaining life.

> *A product stockable unit (SKU) is a sellable product iteration or version under its own orderable part number and price.*

Production Build and Volume Manufacturing

Once the pilot build for the product has passed the first article inspection, internal, and external validation have been completed and most of the defects found have been fixed, product structuring is complete, ordering systems are loaded, demand forecast is in place, and supply chain has been lined up, it is time to kick off the manufacturing belt to put the product into *production*. Production units are basically expansion of the pilot units in mass quantities. In line with the production or volume manufacturing is what is called the *general availability* (GA) of the product, which makes the product available to customers for purchasing. For this reason, the final product is also referred to as the *GA build*. Before the general availability, an *open-books* process has to take place in order to start taking the product orders. We will discuss open-books process under the product launch in later chapters. Once the open-books has been done, the product manager will provide the operations team a target date to *begin-ship*, on which date the product starts shipping to warehouses, partners, and customers.

The manufacturing process and changes in the manufacturing status of the product are controlled through a *manufacturing change order* (MCO) process. Just like an engineering change order process discussed earlier, a manufacturing change order is a controlled process to manage any time a changes is needed to the manufacturing process of the product. Examples of such changes include defects found with certain manufacturing tool or component as well any change in the process itself.

> *A manufacturing change order (MCO) is a controlled change management process to make any changes to the manufacturing process.*

There are two manufacturing models in place in the hi-technology industry these days: the *in-house manufacturing* (IM) and the *contract manufacturing* (CM). The in-house manufacturing is the conventional manufacturing in which a company builds and maintains its own manufacturing facility. Due to facilities and labor costs rising as well as other issues companies have to deal with maintaining a factory such as labor laws and unions, there should be enough overall production volumes for the company to justify the cost of maintaining in-house manufacturing that could be otherwise very expensive. The contract manufacturing, on the other hand, offers competitive manufacturing without dealing with the above-mentioned issues. A contract manufacturer takes the specifications and demand forecast from a vendor company and supplies the manufactured finished product according to the demand. During the past few decades, due to low labor costs and access to cheaper commodities, China, Taiwan, Malaysia, and some other Asian countries have emerged to be as giant contract manufacturing industry. For this reason, many American and European hi-technology companies have outsourced their manufacturing operations to Asia.

Increasingly, a type of contract manufacturing is gaining popularity in the hi-technology industry. It is called the *original design manufacturer* (ODM).

A design manufacturer not only manufactures the product for someone else, but it also provides the complete end-to-end product design and engineering services. During the past decade, many design manufacturers have gone to the level of sophistication where they can handle complex product designs themselves and provide very competitive costs based on their vast supply chain leverage. Many hi-technology companies, particularly in the networking industry, use design manufacturing model combined with the merchant silicon technology to meet the competitive price pressures. Although interesting enough, the detailed discussion of design manufacturing and the manufacturing processes is beyond the scope of this book.

Key Takeaways

The execution phase is the phase when a product really starts taking shape. It is busy time for the functional teams involved in designing, building, and testing the product and for the product manager supervising all activities closely. It is very rewarding for the product manager to see the early samples of the product and then the fully finished product and how a mere dream results into a fully functioning reality. The best part is yet to follow, which is seeing other people wanting to pay for the product.

Remember:

- A business case provides the clear net profit picture by comparing a product's projected revenues and costs and justifies why the product should be built.
- A breakeven point is reached when product has generated enough net revenue equal to the total cost spent on building the product.
- The product life cycle process manages the flow of a product life cycle in a well-defined way by identifying the key deliverables and owners at every stage.
- A product's brand is a unique item such as a product name, a symbol or a logo, or a design that distinguished one vendor's products from another vendor's products of the same type.
- A prototype is an early stage sample product built to test the product functionality and carry forward the on-top product development.

- A beta product is the first manufactured product used for internal and external testing, for regulatory compliances and certifications.
- A pilot product is almost a production grade product that is a candidate final for the shipment and used for last-minute complete checks.
- The first article inspection is an important step at which the product is closely inspected for catching any anomalies when they are first received from the manufacturing facility.
- The marketing forecast is consumed by the operations and manufacturing to determine how many units of product need to be manufactured every quarter.
- Negotiating the costs periodically with existing suppliers, finding alternate sources of supply, and shopping for more efficient components affect the product margins.
- Inventory management is an important activity that needs to be managed ongoing basis.

CHAPTER-4

PRODUCT LAUNCH

The Launch Phase

As the product execution phase comes toward an end and the idea that the product manager had envisioned takes shape of a real and functional product, the preparation to launch the product and letting the world know about it starts. A great product is of no use if no one knows about its existence. Therefore, the product launch involves announcing the product to the parties of interest and letting them know that it is going to be available for sale soon. The product launch therefore includes series of activities and not a single task. It involves spreading the news about the product and actively starting marketing it to build the mindshare. Just like building a space rocket and launching it on a launch pad are two different things, building a product and launching it are two totally different things. A product launch can start several weeks or months before the product actually reaches completion and the *general availability* (GA) milestone and starts shipping.

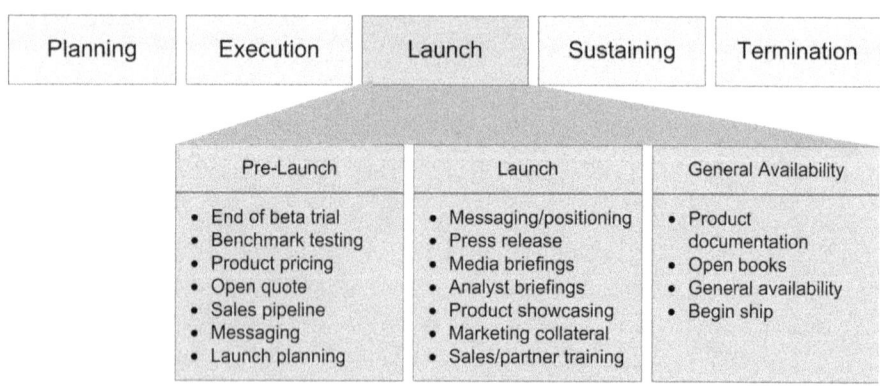

Diagram 1: The Launch Phase

Some of the activities that involve a typical launch readiness plan may include the following activities each of which will be explored in this chapter:

- End of beta trial
- Benchmark testing
- Preliminary pricing
- Open for quote announcement
- Media and analyst briefings
- Customer tours
- Product collateral
- Press release
- Product showcasing
- Open-books announcement
- General availability
- Begin-ship

The type and number of launch activities may vary from product to product and from company to company. We will review multiple tiers of launches later in the chapter. Next, we will explore details of important product launch-related activities.

Prelaunch Activities

End of Beta Trial

As discussed in the previous chapter, the objectives of the beta trial are twofold: first, to test the product in a closer to real environment in which product will eventually be deployed. This is so that timely quality feedback could be collected in terms of customer experience and issues found can be fixed before a wider customer base deploys the product. Second, to let the potential customers know about the product and have them try it, which could result in potential sales. Before the product is launched, it is important to end the beta activities, recall all the beta units, and put a report together based on the lessons learned from the beta trial. This should be done well in advance to incorporate any final touches that could not have been done during the beta trial. Once the beta has ended, the product manager can focus on more involving launch activities.

Benchmark Testing

To build the new product credibility, it is sometimes useful to conduct the product testing through an independent party. The third party can put the product through demanding testing scenarios and endorse its differentiation and performance publically. This could be regarded credible by the potential customers because it is based on a "neutral" party's statements. In addition, most of the time, independent agencies test the product against multiple competing products and given that a product does well comparatively, provides strong evidence for its value proposition. This helps in building the customer confidence in the product and can be an effective sales enablement tool.

In the networking industry, there are several well-known organizations providing competitive testing services such as Lippis, Tolly, University of New Hampshire, Ethernet Alliance Networking Test Committee (EANTC), Network World, Open Networking Forum (ONF), and others. Each of them offers different types of services. Publishing a benchmark report along with the product lunch can draw some good attention to it. Customers then perceive it as a serious product rather than just another product.

Product Pricing

One of the most important tasks that the product manager has to complete before launching the product is *pricing*. Pricing has an extremely important role in the success of a product since in almost every sales opportunity it is one of the most important, if not the topmost, considerations for the end customer's purchasing decision. Pricing the product right pays off for long time. If a product is priced too high, it will result in customers not buying it at all or not buying it much. It will also result in sales teams pushing for very high discount requests under the competitive bidding situations and getting frustrated by losing deals if such is not granted. This also discourages the sales team to quote the product as salespeople are always gravitated toward easier to sell products to retire their quotas. Moreover, high discounts result in poor margins.

If product does not sell in sufficient volumes, eventually its demand forecast will start declining, and it will become very difficult for the commodity team to maintain the required level of costs with the suppliers further eroding the margins. Result—the product will start dying way before its *end of life* stage is due. On the other hand, if a product is priced too low, it may sell in high volumes in certain geographies but with low margins, since there are still discounts involved. In geographies where high discounts are common regardless of the list price, margins will be poor because customers perceive value through how much discount was offered to them. The end result is poor profitability from the product. It is therefore important that product is priced just right, keeping the competitive landscape, different geographies, and buying behaviors in mind. We will explore these tactics in more details later.

Average Sale Price

As we had discussed earlier during the planning phase, it is almost nonexistent in the networking industry that a product sells at the actual published price or the *list price*. There is always a *discount* expected by the customer. The customers have interesting ways to command the discounts that could vary based on the region, deal size, customer brand name, competitive bidding, and other factors. Once the estimated list price has been determined per unit, the product manager will also estimate the average *discount* per unit that it expects to allow. Different industries and products have different discount levels, and it also varies based on the geography. The actual price at which a product end up selling most of the time, after the average discount has been applied, is called the product's *average sale price* (ASP) and is calculated as below:

Average Sale Price (ASP) = Product List Price X (100% – % Product Discount)

If the product pricing or discount structure is not competitive, there will be lots of high discount requests coming from the sales and the partners to the product manager for approval. This is referred to as *nonstandard pricing* (NSP) requests. A high number of NSP requests indicate that the product pricing is no more competitive or discount limit needs to be raised, and it is time for the product manager to recalibrate the pricing and the discount structure as needed.

Regional Pricing and Discount Structure

One important factor to consider when pricing a product is demographics and cultural behaviors. For example, in North America, people like to know clearly upfront what a product is going to cost them after all discounts and rebates are applied, so the product average sale price is important. In Asian countries, on the other hand, people give lot more attention to the discount in addition to the average sale price; and if the discount is higher, the perception usually is that it is a good deal. Therefore, the product manager needs to price the product accordingly to compensate for higher regional discounts and still achieve the target margins. In this case, the product manager may need to *uplift* the pricing by certain percentage in certain regions and publish multiple price lists for multiple geographies. This is in fact a common practice.

Sometimes there are specific discounts granted to the sales teams or channel partners per the nonstandard pricing for an end-user deal through a preapproved *discount authorization* (DA). There is a unique discount authorization number associated with every such deal maintained in the sales quote tool. This gives the channel partners freedom to apply the discount when and where they need to. The vendor company tracks it through the *point-of-sale* (PoS) data it receives from the channel partners to issue the credit back to the channel partners according to the applied discount.

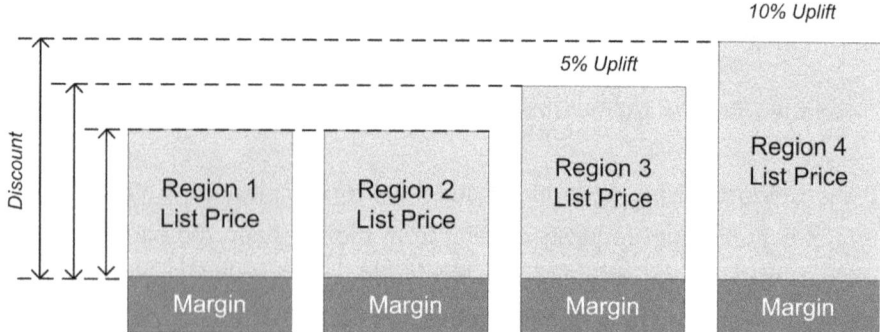

Diagram 2: Regional Pricing Strategy

Pricing Methodology

The pricing methodology can vary from industry to industry, but there are some commonalities and fundamentals to it. First set of data point that a product manager should have in hand is the pricing of similar competitive products at the given time. There are legal ways to obtain the competitive list pricing, such as through a public bidding deal, competitor's public announcements and press releases, and other such sources. Some competitors may not hide list prices at all and may have published on their websites.

The first method of pricing based on the competitive benchmarking is pioneered by the author and referred to as the *three-dimensional (3D) pricing strategy*. Per this method, once the product manager has competitive pricing of similar class of products, it can be ranked in an order from the lowest to the highest. This first set of ranking marks the lowest (floor) and the highest (ceiling) reference price points, in between of which the product could be priced somewhere. As another reference point, the closets target products should be ranked in the order of importance such that the topmost target competitive product that needs to be competed against should be on the top and so forth. This produces a second set of ranking for reference. A third set of ranking is useful to have as well. In this case, the product manager ranks the competitive products in the order of strongest in terms of feature richness, performance, scale or whatever attributes matter, and so forth.

Diagram 3: 3D Competitive Product Ranking

With these three stacks, the product manager has a pretty good view of the highest- and the lowest-priced competition, the most important and the least important competition, and the strongest and the weakest competition. Keeping the differentiation, strengths, and weaknesses in mind, the product manager then relatively inserts the new product in the three stacks. This produces three vectors that point to where the product should be priced at loosely. The product manager takes the three baseline prices and ranks them in the order of what matters for the product positioning and messaging the most. That is, if the product will be primarily marketed based on the differentiation, then the baseline price out of the strength-based stack comes on top. The product manager now has the final low and high range of the target price and can pick a price that fits with the marketing strategy or simply take the median. This pricing strategy is a *top-down* approach and helps with the top-line revenue growth by being within the competitive bidding range almost always.

To be sure about which price to pick, having a margin view as a second data point helps. In this *bottom-up* approach, the product manager sets the pricing purely based on target margin. That is, the product manager takes the burdened-COGS and sets a list price, which at the target discount generates the desired margin. To be precise, two baseline prices can be calculated. One based on the highest possible, but usually undesirable discount, and other based on average expected discount. The final price can be picked based on the median of the two prices. This pricing strategy helps in assuring the bottom-line growth and profitability, although it may not always be competitive. As the very final step, once the product manager has derived the prices based on the two methods above, final product price can be determined as a median of the two. The process is repeated for each product SKU.

EXAMPLE 1

Pricing Calculation: 10G ToR Fiber Model

- First using the 3D top-down pricing method and ranking competitive products:

Price Based Ranking	Target Based Ranking	Strength Based Ranking
Product A ($25,000)	Product B ($22,000)	Product A ($25,000)
Product B ($22,000)	Product C ($18,000)	Product B ($22,000)
Product C ($18,000)	Product A ($25,000)	Product C ($18,000)
Product D ($15,000)	Product D ($15,000)	Product D ($15,000)

- Inserting the new product relatively into the three stacks:

Price Based Ranking	Target Based Ranking	Strength Based Ranking
Product A ($25,000)	Product B ($22,000)	New Product ($25,000)
Product B ($22,000)	New Product ($21,000)	Product A ($25,000)
New Product ($19,000)	Product C ($18,000)	Product B ($22,000)
Product C ($18,000)	Product A ($25,000)	Product C ($18,000)
Product D ($15,000)	Product D ($15,000)	Product D ($15,000)

- This provides three base line prices that are ranked in the order of importance:
  - o $21,000—*Generally competitive*
  - o $19,000—*Aggressive for competition takeout*
  - o $25,000—*Premium for differentiation*

- Instead of picking a median, we pick a price in between the first two to yield the most desirable competitive pricing since last price is farther from most of the competitive prices.

- Next, using the margin based bottom-up method with:
  - Highest discount = 70%
  - Average discount = 60%
  - Target minimum desired margin = 55%

	Target List Price	Target ASP	Transfer COGS	Burdened CCOGS @ 25% OH	Margin @ ASP
Price at 70% discount	$23,500	$7,050	$2,500	$3,125	55%
Price at 60% discount	$18,000	$7,200	$2,500	$3,125	56%

- The median of the two prices comes out to be $20,750, and we set the price as $20,000 in this case as well. This provides competitive price advantage while good margin at average discount.

- This price is then adjusted for multiple geographic theaters based on average or above-average discount needs. Therefore, the final regional product prices are:
  - North America = $20,000
  - Latin America = $20,000
  - Europe (10% uplift) = $22,000
  - Asia Pacific (20% uplift) = $24,000

Once the pricing has been finalized, a *pricing proposal* is put together by the product manager for the executive team's or a *pricing committee's* review. The pricing proposal includes the competitive pricing, relative positioning, and margin analysis on per product SKU basis. In addition, the proposal usually includes the *aggregate margin analysis* for typical expected *configurations* of the product in which more than one product component will be sold together. Once the product pricing has been approved by the pricing committee, it is ready for publishing and for the sales team's and the partner's use. However, the actual pricing is not published to anyone until the open-books stage is reached that we will discuss later in the chapter.

Price Premium

Sometimes, it is possible to command a premium on price that may be higher than the usual competitive pricing, if:

- the product is unique in type with little or no competition.
- the company or the product is well known, and the product can be differentiated with a wider gap from the competitive products and customers would pay for the additional value that matters to them.
- the product is highly sought out based on customer desirability and need.
- the company brand name is of significant value to charge a premium for because it associates with better quality and customer experience in customer's minds.

Under those situations, customers could be charged higher for the product with reference to the competitive offerings, and most likely they would pay the premium. It is not hard to find many examples around us that it is indeed pretty common. However, charging premium under the wrong combination of things can backfire. For example, chances are slim that customers will be willing to pay a premium for a product that may be better than the competitive products, but the company and the product are new to the market. The relationship of brand name, differentiation, and customer perception is an interesting one and a long discussion in itself. The customer perception is not a result of a single incident or a single product experience. It takes hard work and a successful history of positive experiences.

Preliminary Price

Before the product manager can work out detailed and final pricing for the product, sales and partners may need it for responding to the customer queries and generating sales estimates. Even if the product manager has the actual price finalized, it may not be strategically advantageous to expose it way before the product is launched. The competition may learn from it and may come up with a better pricing model, which could hurt the new product business. When it comes to the pricing, the surprise factor is a good competitive tool. For this reason, usually a *not-to-exceed* (NTE) price is used as the preliminary pricing. The not-to-exceed price is an inflated list price that is advertised as the maximum possible price a customer would have to pay. The actual price in the product manager's mind is somewhere below the not-to-exceed price. It is inflated to add the safety room for any upward price adjustment needed before the general availability of the product, in case a new surprise product announcement from the competition or any other market disruption changes the competitive landscape.

The not-to-exceed price is published to the sales and the partners for issuing early sales quotes and is a commitment that the actual product price will never exceed the published price, but may be equal or lower. The not-to-exceed price is only used until the actual list price is available.

The maximum ceiling is set for the not-to-exceed price as an assurance and aid to the customers so that they can plan their spending budgets in advance knowing that at most they will need to pay according to the maximum price advertised (minus discounts) and not higher than that. This adds the predictability to the picture. The product manager should set the not-to-exceed price such that it is higher enough from the actual list price to keep room for upward adjustments, but at the same time, not so high compared to the competitive offerings that it may prevent the customers from even considering the product. Usually, the not-to-exceed price can be established and uploaded into the sales tools by the *open-quote* stage.

> *The not-to-exceed (NTE) price is a preliminary price issued to sales and partners for quoting the product before it is generally available, with a commitment that the actual product price will not exceed it.*

EXAMPLE 2

Regional NTE Pricing: 10G ToR Fiber Model

Region	Expected List Price	NTE List Price	Uplift
North America	$20,000	$24,000	20%
Latin America	$20,000	$24,000	20%
Europe	$22,000	$27,500	22%
Asia Pacific	$24,000	$29,000	20%

Open for Quote

As soon as the product starts getting ready for completion, it is time for the sales team and channel partners to start "pitching" the product to the potential customers in order to create sales opportunities. Before this can happen, the preliminary product list price must be set and communicated to the sales team. Once the not-to-exceed price is determined, the product manager opens up the product for the sales team and partners to start quoting to the potential customers. This stage is called *open for quote* or just *open-quote*.

> The open-quote (OQ) stage for a product indicates to sales and partners that the product orderable part numbers and the preliminary pricing is now available for them to issue sales quotes.

For the open-quote, the product manager follows a process similar to that for the bill-of-material request discussed earlier. The process consists of submitting all product stock keeping units to the product data management team with the not-to-exceed pricing information so that the pricing along with the orderable marketing part numbers can be loaded into the sales quote tool, such as *Siebel* or *Sales Force*. The data management team loads the information into the quote tool by the target completion date that the product manager provides. The product manager also issues an open-quote announcement to the sales and partner community to let them know that the product is now available for quoting. This is a turning point in the product life since now sales teams can start building what is referred to as a *sales pipeline*—a stacked and ranked queue of potential sales opportunities that should start growing. The sales pipeline is extremely important for a variety of reasons and is discussed next.

EXAMPLE 3

Open-Quote Request

<table>
<tr><td colspan="11" align="center">Open-Quote Request</td></tr>
<tr>
<td colspan="2">Number:
ECO-123</td>
<td>Business Model:
Z-Series</td>
<td colspan="2">Requestor:
Nadeem Zahid</td>
<td>Requestor's Job Role:
Product Manager</td>
<td>Request Date:
15/10/2013</td>
<td colspan="4">Target Complete Date:
30/10/2013</td>
</tr>
<tr>
<td>Marketing PN</td>
<td>Manu-facturing PN</td>
<td>Product SKU</td>
<td></td>
<td></td>
<td></td>
<td>Revenue Treatment</td>
<td>NTE List Price</td>
<td colspan="2">Warranty</td>
</tr>
<tr>
<td></td>
<td></td>
<td></td>
<td>Long Description</td>
<td>Product Line</td>
<td>Product Category</td>
<td></td>
<td></td>
<td>HW</td>
<td>SW</td>
</tr>
<tr>
<td>101</td>
<td>300-1111</td>
<td>TOR-10G-48F</td>
<td>48-port 10GbE SFP+ ToR Switch</td>
<td>Z-Series</td>
<td>Switch</td>
<td>Relative Sell Price</td>
<td>$24,000</td>
<td>1 year</td>
<td>90 days</td>
</tr>
<tr>
<td>102</td>
<td>300-1112</td>
<td>TOR-10G-48C</td>
<td>48-port 10GbE RJ45 ToR Switch</td>
<td>Z-Series</td>
<td>Switch</td>
<td>Relative Sell Price</td>
<td>$30,000</td>
<td>1 year</td>
<td>90 days</td>
</tr>
<tr>
<td>103</td>
<td>300-1113</td>
<td>TOR-UL-40G-4</td>
<td>4-port 40GbE Uplink Module</td>
<td>Z-Series</td>
<td>Accessory</td>
<td>Relative Sell Price</td>
<td>$5,000</td>
<td>1 year</td>
<td>90 days</td>
</tr>
<tr>
<td>104</td>
<td>300-1114</td>
<td>TOR-STACK-480</td>
<td>480Gbps Stacking Module</td>
<td>Z-Series</td>
<td>Accessory</td>
<td>Relative Sell Price</td>
<td>$2,500</td>
<td>1 year</td>
<td>90 days</td>
</tr>
<tr>
<td>105</td>
<td>300-1115</td>
<td>TOR-L3-LIC</td>
<td>Layer 3 Feature License</td>
<td>Z-Series</td>
<td>License</td>
<td>Residual</td>
<td>$6,000</td>
<td>N/A</td>
<td>90 days</td>
</tr>
</table>

Sales Pipeline

Once the product is open for sales teams to quote, they start registering the sales opportunities into a system or tool maintained for this purpose. When a salesperson registers a sales opportunity, usually per the process, it also assigns a *probability level* (such as 75 percent or 50 percent) to each opportunity, indicating the likelihood or the confidence level if a quote will actually translate into an actual *booking* or an order. The changes in the probability levels indicate how a given *deal* is progressing. As more and more quotes are registered into the system, this starts building up a *sales opportunity pipeline.* The sales pipeline now looks like a long list or queue in which opportunities are relatively ranked.

> A sales pipeline consists of the potential sales opportunities at the time, ranked in terms of different probability levels for turning into actual bookings or orders.

The product manager may monitor the sales pipeline closely and on weekly basis to ensure that the product is having good traction in the market, given the access to the sales data. A healthy product should have a growing pipeline, and a significant portion of the pipeline should translate into bookings, meaning actual sales. We will explore sales pipeline in detail under the sustaining phase. Once there is a healthy rate of sales pipeline buildup after the open-quote, the product manager can start tuning the product demand forecast more accurately as preparation for the product launch.

> A booking occurs when a customer commits to buy a product. A booking is the promise to revenue.

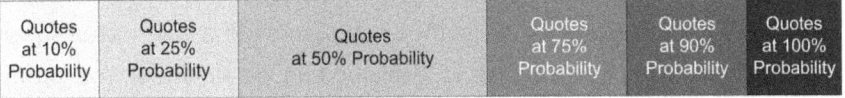

Diagram 4: Sales Pipeline Example

Product Messaging

One item that serves as the foundation for almost all marketing material about a product is the *messaging*. The product messaging conveys the *key message* about a product and its key strengths that the product manager wants people to remember and associate with it. Messaging concentrates the most important points about the product into a concentrated and concise form. Developing good messaging is more art than science and possesses the following qualities:

- It is specific and provides concrete and quantifiable facts about the product

- It links the technological innovation and strengths with business advantages
- It is simple and easy to understand

The quantifiable claims made in the product messaging should be backed up through benchmark testing or other verifiable data points such as competitive data sheets.

> *The product messaging conveys the key message about a product, and its key strengths that are intended to be remembered and associated with the product.*

Without the messaging, there is not much to be communicated. Messaging is required to highlight the clear differentiation against competition, which means highlighting why and how new product is better and how much better. Good messaging starts with a good *messaging framework*, which links apparently heterogeneous attributes together so that connections among them could be established in a sensible and cohesive way. A good messaging framework generates messaging for different levels of audience or readers in a customer organization. The author has pioneered a method of *layered messaging framework* (LMF) that is explained in detail under the product marketing chapter later in the book.

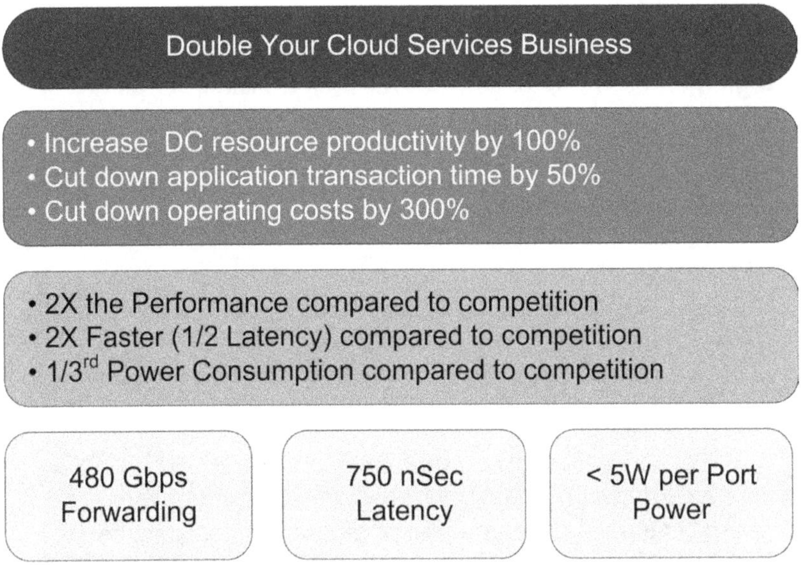

Double Your Cloud Services Business

- Increase DC resource productivity by 100%
- Cut down application transaction time by 50%
- Cut down operating costs by 300%

- 2X the Performance compared to competition
- 2X Faster (1/2 Latency) compared to competition
- 1/3rd Power Consumption compared to competition

| 480 Gbps Forwarding | 750 nSec Latency | < 5W per Port Power |

Diagram 5: Layered Messaging Framework

Having the product messaging nailed down before the product launch is critical so that the marketing material can be based on it and can expand it into detailed value proposition throughout the content. For this reason, there should be formally a *messaging and positioning guide* document drafted by the marketing team that can be referenced for developing the marketing collateral as well as to train sales and partners. Messaging guide provides a common denominator, which results in consistency in articulating the product differentiation and value proposition no matter which piece of the marketing collateral a potential customer or analyst refers to. This suppresses multiple stories to be told that may not agree with each other. While the product manager may not be the originator of the marketing material, it must be the key party for providing the input into the messaging framework. Messaging developed without product manager's input is commonly found to be ungrounded, weaker, and factually off since no one else understands the product better.

EXAMPLE 4

EXAMPLE 4

Messaging Framework: 10GbE ToR for a Cloud Data Center

Doubles Your Cloud Services Business

- Increases DC resource productivity by 100%
- Cuts down application transaction time by 50%
- Cuts down operating costs by 300%

- 2X the Performance compared to competition
- 2X Faster (1/2 Latency) compared to competition
- $1/3^{rd}$ Power Consumption compared to competition

| 480 Gbps Forwarding | 750 nSec Latency | < 5W per Port Power |

Product Launch Planning

Building a product is not enough by itself. The target customers must find out about what the product is and what it is capable of doing in order to buy it. Therefore, as the product is getting ready, the first step is to start building the awareness about it. This is marketing 101. If there is no awareness, there is no marketing, and there are no sales, and therefore there is no business. The awareness building starts with the product launch and continues as long as the product lives. The product launch planning is needed to appropriately plan and successfully execute upon a launch. A number of resources and significant money is usually invested in building a product. It is important that the required due diligence and preparation is done for its launch as well to make the required level of impact in the market and to get the needed customer attention. The product launch is not a one-time event as commonly misunderstood, but rather a series of events that could last for weeks or even months. A typical product launch begins with a public announcement about the product and ends with the general availability of the product. There are several events that occur in between, depending on the type of industry, demographics, product, and marketing funds available. The product launch typically includes activities like announcing the availability of the product, creating and propagating product messaging and differentiation, building and distributing product collateral, conducting media and analyst briefings, launching product web page, showcasing the product at major public events.

> *The product launch involves announcing the product internally and externally and to the potential customers to build awareness as well as providing necessary marketing content to enable the product sales.*

Launch Activity	Level 1	Level 2	Level 3
• Announce product at industry event	●	○	○
• Press release	●	○	○
• Sales & Partner email flash	●	●	●
• Customer email blast	●	●	●
• Web dev updates	●	●	●
• Social media (facebook, twitter)	●	○	◐
• Geographic customization	●	○	○
• Press & Industry Analyst briefings	●	○	○
• Internal webcasts	●	●	◐
• Positioning and Messaging doc	●	●	○
• Data sheet update	●	●	◐
• FAQ	●	●	○
• Product landing page	●	●	●
• Customer presentation	●	●	●
• Technical presentation	●	◐	○
• Competitive marketing	●	◐	○
• Sales product training	●	●	○
• Partner product training	●	●	○

Diagram 6: Product Launch Plan

The diagram above shows a way of planning different levels of launches and what sort of activities would apply under those situations. A company can build and maintain such a master launch plan that can be applied to product launches as needed. In this case, multiple levels of launches are created as a master guideline. When a product is getting ready for the launch, appropriate launch level is determined applicable to the product, which then dictates what events would take place for the launch, which resources, and how much marketing budget will be needed to accomplish those. Per this template, the emphasis is on building both external and internal awareness about the product. Internal awareness is important as well since the sales and business development teams need to be trained and armed with the sales enablement content and messaging in order to sell the product. In most companies that take product marketing and evangelism seriously, the product launches are managed by a dedicated *product marketing* team under a focused role of a *product marketing manager* (PMM). A marketing manager is responsible for promoting and evangelizing the products and providing them the due attention throughout their life. In this capacity, the marketing manager works closely with the product manager, sales, and partners.

In many cases, there is also a *launch project manager* (LPM) who plans and coordinates all launch activities among multiple teams involved. The project manager and the marketing manager generally do not engage in the concept or early execution phases of the product life cycle but rather during late execution phase, sometime before the launch in order to start planning. In smaller companies, sometimes the product manager performs the dual role of being a product manager as well as a marketing manager and conducting the product launch. We will explore this aspect more later on under the product marketing chapter. However, in this case, it is very hard to do a level-1 launch because the product manager's core role of building new products is already a rigorous one. Regardless of who performs the product marketing activities, the launch process and its purpose stay the same.

Launch Activities

Messaging and Positioning Guide
As discussed earlier, the product messaging serves as the foundation for almost all the product marketing material and for the external presentations. The first most important task of any serious product launch is completion of a *messaging and positioning guide* drafted by the product marketing team, which can be referenced for developing the marketing collateral as well as

to train the sales and the partners. The messaging guide provides a common denominator for consistency in articulating the product differentiation and value proposition no matter which piece of marketing collateral a potential customer or analyst refers to.

Press Release

Usually, the product launch begins with a public announcement through a *press release* (PR). A company's *analyst and public relations* (AR/PR) team, which is part of the marketing organization and is responsible for preparing and issuing the press releases. A press release announces the new product or solution availability and provides its introduction, explains what industry pain points it solves and how it solves them better than the competition, and usually ends with the introductory pricing and target general availability time frame of the product. The press release usually includes endorsements and quotes from one or more industry renowned personalities for brand elevation and credibility building.

Although the press release is generally not written by the product manager, however, it is recommended that if it involves a product announcement, it should be reviewed by the product manager and marketing. It helps avoid any technical mistakes that could prove to be embarrassing later on. Commonly, the timing of the press release is lined up with a major industry event to get the maximum customer and media attention, such as a trade show or public conference. The press release is usually followed by product showcase and media and analysts briefings at the event. Therefore, the press release is usually the trigger of a series of planned events that take place as part of the product launch. The company's investors and shareholders closely monitor the press releases, and it acts as one of the indicators of company's healthy activity in terms of innovation and future prospect.

EXAMPLE 5

Press Release

Nubes Networks Handles the Increasing 10 GbE Demand in Data Centers

Las Vegas/March 15, 2014/PRNewswire/—Nubes Networks today announced new Z-series 10 Gigabit Ethernet Top-of-Rack switches for data center server and storage connectivity that are fully controllable through Software Defined Networking (SDN). The product will be showcased at the Interop, Las Vegas trade show on booth number 123.

Per mainstream market research companies, 10 GbE will become the major connectivity option for the data center and cloud servers in the next few years. This will be driven by the LAN on motherboard (LoM) based server mass production expected in next couple of years coupled with data center storage convergence of traditional LAN and SAN over same Ethernet network.

"Data Centers are going through a major transition fueled by cloud services, virtualization, and total cost of ownership optimization. This demands an efficient, high bandwidth, yet cost-effective networking fabric. Nubes Networks new 10 GbE switches help cloud operators meet the above needs in most cost-effective way without compromising on network performance," said Nadeem Zahid, Director of Product Management at Nubes Networks. *"Furthermore, SDN is changing the paradigm how data center networks will be managed and operated from a unified command and control perspective with plug-n-play open networking. Nubes Networks Z-series leads the wave in these respects as well."*

New 10 GbE Z-series switches are available in both fiber and copper connectivity options. With 480 Gbps wire-speed performance and 40 GbE uplink connectivity. Z-series offers next-generation ToR solution that can scale up to 8 units in a single stack that can be managed as a single virtual switch. New Z-series will start shipping during 2nd half of 2014 with a starting US list price of $24,000.

Media and Analyst Briefings

As the product announcement is on the way out, the AR/PR team starts lining up the media and industry analyst briefings regarding the new product introduction. One of the most important launch activities is the analyst and media briefings. These briefings are opportunities for a product or marketing manager to tell the product or solution "story" in a way it should be conveyed. The product manager or the marketing manager is usually the leading personality providing the product overview and differentiation to the media and analysis during the briefings. The media and analyst companies then publish articles with their own views and analysis about the new product. This passes through large number of readers' eyes and generates good product awareness.

The primary purpose of media and analyst briefings is to generate the positive perception and influence regarding new product in the industry. A good practice is to always seek feedback on the spot from the media and analysts on what they think about the new product, its value proposition, and market impact. This feedback can be a valuable source for the marketing teams to refine the product messaging and content. We will discuss more on this chapter under the product marketing topic later in the book.

Product Showcasing

Usually, the press release announcing the new product is aligned with a major trade show or industry event. The product marketing usually plans for presence at major trade show events by placing a booth or having a public speaking opportunity. This provides an opportunity for a company to display its latest and greatest products and solutions to thousands of visitors including potential customers, partners, media, and analysts. Interested customers stop by at the booth, attracted by free giveaways or other incentives, and ask questions. They are usually provided with the product collateral for study, while their contact information is captured as potential sales leads for later use. Those leads are reviewed, scored, and followed with by the sale representatives to generate new business.

Showcasing the product is not limited to the trade shows only. The product marketing team also showcases the products in the company's executive briefing center (EBC), which is maintained at the company's own location for inviting the customer executives and decision makers. The executive briefing center focuses on providing the corporate, products, and solutions overview to the customer executives who the sales team brings in for the visit. The product manager, product marketing, and the sales teams are usually the key participants in customer briefings. The visiting customers are presented with the product overview followed by a tour of the product showcase area where the company products are displayed. Executive briefings generally result into rather long term sales returns.

Yet another way to showcase the newly launched product is to reserve certain number of product units for the sales team's and the partners' use, which can be distributed worldwide to become part of the localized *product demonstration pools (demo pools)*. The sales engineering team maintains those demo pools and rotates the product around for showcasing and demonstrating it to the potential customers through a *request-and-reserve* process. This provides a personalized experience to potential customers. The new product can also be part of a *try and buy program* in which case a customer has the opportunity to try the product in its environment before buying it. The try and buy works well for very expensive products where customers may be reluctant to trust the new product and aren't willing to pay for it without having complete satisfaction about it.

Another activity that may or may not always happen as part of the launch is a planned customer tour. The marketing team selects important customers who are on the "must win" list of the company or are existing customers and plans regional and multicity tours visiting them at their locations to build the product and brand awareness. The product manager may or may not accompany the tour, and it is generally managed by the marketing teams. Although a customer tour may not translate into immediate sales pipeline necessarily, it eventually pays off in the long run. A similar strategy is used with the channel partners. The customer tours can be planned jointly with the partners, and in this way, the partners are also educated about the new product and how they can help push the new product sales.

Product Collateral

Out of all things important, the product collateral has its own significance and is most sought out pieces of information by the customers during the launch phase as well as throughout the product life. Whenever a potential customer needs information about a company's products or services, they look for the product collateral on the company website or request hard copies from a sales representative or partners. There are, in fact, several types of product collateral focused on variety of objectives and target readers. Just like the product messaging, the collateral has to be focused on who the reader will be. For the networking products, some of the common product collateral available at the product launch, depending on the level of the launch, may include:

- Product landing web page
- Product brief
- Product data sheet
- Benchmark test report
- Product videos
- Flash animations
- Solution briefs
- Technical briefs
- Customer presentation
- Technical presentation
- Social media
- Email news blasts

We will review most of the above product collateral pieces in detail under the product marketing chapter later on.

The *landing web page* serves the most important purpose in providing firsthand information about a product. Many technology companies have multiple paths to find the product information through the product landing page. The product web page serves as the central place for holding important information for customers and partners, such as the product overview, specifications, datasheet, solution briefs, white papers, and other documentation. A *product brief* is an overall product guide that provides a brief overview of overall product portfolio and offerings of a company. Sometimes a *quick reference guide* (QRG) is used to provide a brief side-by-side overview of the products and their specifications. A product brief is usually useful as the firsthand overview for customers not yet fully familiar with the products and offerings of a company.

The *data sheet* is viewed as the most authentic and credible source of product information by the customers. As the name suggests, the data sheet collects the most important data about the product in a single document. The data sheet generally consists of the product overview, key highlights, value proposition, detailed product specifications, any industry standards compliance, and ordering information. A *benchmark test report* helps build product credibility. Benchmark testing is done through an independent party to qualify the product strengths and features. The benchmark testing can be expensive but its impact can last for long time.

A *video data sheet* is a few minutes long video in which the product manager talks about the key features and capabilities of the product. This attracts more customers to learn about the product compared to traditional paper data sheets, and it complements the paper data sheet rather than replacing it. In addition to the video data sheet, there can be other types of videos developed for marketing the product. For example, videos of product showcasing at the industry events, videos from any early customer tours, interview-style interactive videos with company executives and key industry analysts, and customer testimonials about the product. In addition to the videos, interest in animated video-style illustrations is also on the rise. These *flash animations* can be a great marketing tool in the hi-technology world since there are usually complex product- and solutions-related concepts that need to be presented in a simplest possible way to the potential customers.

A *solution brief* is usually a one- or two-page document focused on a particular industry vertical based solution and application of the product. It explains the key requirements of the solution and how the product fits those requirements within the overall solution. The solution briefs are popular with customers as they are short and they provide the insight into much more relative real applications for them. A *technical brief* is also a short and focused document that is focused on a particular technology and how it helps solve certain challenges. It can also be focused on a set of technologies that work and get deployed together, as a meaningful solution.

A *customer-facing presentation* consists of set of PowerPoint slides that is used to pitch the product to customers. It is a product overview with key messaging and differentiation highlighted, which sales team and partners can leverage to present to potential customers. The customer presentations are used routinely in the executive briefing center for customer meetings. A *technical presentation* consists of set of PowerPoint slides that explain the product architecture, technical features, and functionalities in detail. The technical presentation is focused to cater more technical audience on the customer side. It is also used internally for training the technical resources such as the technical support and the technical marketing teams.

With the explosion of *social media* such as Facebook, Twitter, LinkedIn, and rest, more and more people are spending time online to seek and share information. This opens up a huge opportunity to market the new product online and reach masses of people simultaneously. The product marketing team can put a page on Facebook to share the news, industry events, customer testimonials, products and solutions videos, and other interesting content to spread the word. The customers and the partners can comment on it, like it, and share it, starting an interactive sequence of feedback and response. The product marketing team could tweet important news and highlights as well throughout the launch. Additionally, putting up and participating on well-known blogs online is a good strategy. Advertising on the industry-specific popular web pages and blogs where targeted customers come and read information is another marketing strategy. It helps build brand and familiarize a company and its products to those who may not know about it.

As the new product is launched, sending out *email flash announcement* to existing and potential customers and partners can get their attention towards the new product. The art of successful email news flash is that it should be relevant, concise, and attractive. Since these days, email tools allow embedding graphics and videos inside the messages; some really influential messages can be created and broadcasted to the target customers.

Sales and Partner Training

One of the most important activities before launching the product is to train the sales force and the partners on the new product. Generally, this "how to sell" training focuses on the product features, differentiation, positioning, applications, and key messaging as well as any weaker points to tackle if attacked by the competition. The product manager is the key participant in the training usually and partners with the rest of the marketing and sales resources to structure and deliver the training. The customer- and partner-facing product update and training could be on-site or off-site. It could be organized at leading partner or company locations where partners or customers could be invited to attend. Or alternatively, it could be virtual training in the form of webinars that anyone could conveniently attend from anywhere.

Generally, a good training is segmented in terms of the content focused at two different sets of audience. One part is focused on the salespeople such as the account managers (AM), channel account managers (CAM), business development manager (BDM), and other sales representatives. This part of the training focuses on high-level product pitch, positioning, market opportunity and objectives, key messaging, value proposition, pricing, commercial options, etc. The other part focuses on the sales engineering (SE) and technical resources that in turn support the salespeople. This part focuses on deep dive product details such as product architecture, features, use cases, configurations, competitive analysis, etc. Furthermore, the training content should be segmented for the sales team and that for the partners since not every detail could be shared with the partners; and on top of that, partners have some unique training requirements. We will explore more on this topic later in the book under the product marketing chapter.

General Availability

Product Documentation

Besides more marketing-focused collateral discussed above, there is also some basic but important product documentation that must be available by the product launch. For the hardware products such as the networking switches or routers, there is usually a *hardware install guide* that provides in depth details about the product hardware specifications, components, installation instructions, operating conditions and procedures, electrical and mechanical characteristics, etc. Hardware install guide is the first product document that is consulted by the network engineers on the customer site after purchasing the product to have it installed and to put it into production.

There is usually also a *software configuration guide* or *software concept guide* that provides in depth details about the software features and functionality supported on the product. It also provides the *command line interface* (CLI) and how to configure the features and elaborates any caveats or limitations. Unlike the hardware guide, that does not change much over time; the software guide is updated, along with the *release notes* (RN), every time a new software release is made available. The release notes is a short document that ships with every new hardware or software release and lists what is new and what are the important items that the customer must be aware of, such as the fixed and outstanding (not fixed) defects, any caveats, and the proposed work-around.

Open-Books and General Availability

As the product launch is started, the product manager works with the core team to have the product ready for its *general availability* (GA) and shipping. The general availability means that the product is now available to be sold and shipped to anyone openly. A product could also be made *limited available* (LA) before being made generally available. The limited availability means that the product is being made available to only certain select customers or in certain geographies only and not to everyone or everywhere. The limited availability milestone is more often seen with the software products than with the hardware products.

> *The general availability (GA) of a product means that the product is now available to be sold and shipped to anyone openly.*

Several things must happen before a product could be declared as generally available:

- All product hardware and software development and testing must be complete.
- The product must have passed the first article inspection. As discussed earlier, the product and its packaging are fully inspected for expected customer experience. Any issues found should be resolved before the general availability.
- All critical software defects must have been fixed and software should be stable enough to ensure good customer experience. This is done through a software defect triage process where *must-fix issues* are isolated from the rest of the issues, and these issues are considered *showstoppers* for the general availability release until fixed. This is discussed more under the sustaining phase later.
- All product documentation referred to as the *technical publications* should have been completed, including the hardware install guide, software configuration guide, release notes, and any documentation that must be shipped with the product on a CD or in a printed form inside the product packaging. Any translations into regional languages should have been completed.
- All product safety and compliance testing per the regulatory agency requirements should have been completed and reports obtained. The product labels indicate the conformance to those standards. It should be assured that wherever the product is intended to be sold, those country-specific regulatory issues under *global trade and compliance (GTC)* rules have been taken care of. Otherwise, importing the product into a country could be considered illegal and may result in legal issues.
- The demand forecast should have been completed. This ensures that adequate supply of components required for manufacturing the product has been secured, and enough product units can

be manufactured and stocked to meet the customer lead time expectations.

- The ordering and fulfillment system and order processing flow should have been tested to make sure that when real orders start coming, system can accept and track orders as expected, and the product can be shipped against those orders.

One last important task that the product manager has to complete is called *opening the books* or *open-books*. Open-books means, in financial terms, opening the accounting books for the product revenue recognition. At this step, the product's final pricing is uploaded into the ordering system, and it is ready to accept the real orders. Just like the bill-of-material request or the open-quote request, the product manager submits the request for open-books to the product data team with the target completion date. An open-books announcement is also sent out to the sales and partners, indicating the date when orders can be submitted and product can ship. The announcement should be sent early enough to provide the partners opportunity to upload the pricing information into their own ordering systems and test it.

> The open-books (OB) means in financial terms opening the accounting books for the product revenue recognition.

As the final step, called a *begin-ship*, the product manager provides a target date to the operations team when the product should start shipping to the customers from the warehouses. The operations team makes sure that they have moved sufficient product inventory in the warehouses worldwide and to the distributors in anticipation of forecasted shipments. Once the product starts bookings on the promised date, it is now general available for anyone willing to buy the product.

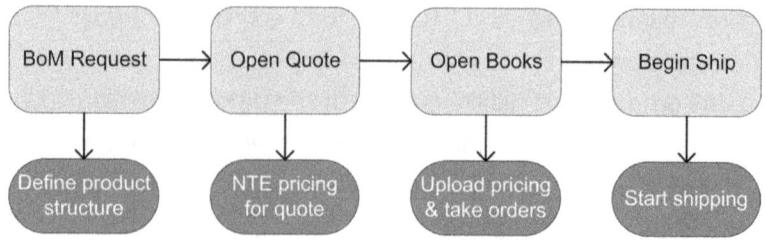

Diagram 7: Different Stages of Product Readiness for GA

Key Takeaways

The product launch is an exciting time for the product manager. The idea that turned into a real product now gets under the spotlight. Everyone in the marketing organization works to get the product its due attention. The curtain is removed, and the outside world can now take a look and get amazed. Questions get asked filled with curiosity and the value proposition, and differentiation is highlighted based on the key messaging. The product is ready for some action.

Remember:

- A great product is of no use if no one knows about its existence.
- The product launch involves announcing the product internally and externally to potential customers to build awareness and enable sales.
- The product messaging conveys the key message about a product and its key strengths that are intended to be remembered and associated with it.
- An open-quote stage for the product indicates to sales team and partners that the product and its preliminary pricing is now available for them to issue sales quotes.
- A sales pipeline consists of all potential sales opportunities ranked in terms of different probability for a possible booking.
- If a product is priced too high, it will result in customers not buying it at all or not buying it much. If it is priced too low, it may result in poor margins.

- The not-to-exceed price is a preliminary price issued to sales for quoting the product before it is generally available.
- The open-books means opening the accounting books for the product revenue recognition.
- The general availability of a product means that product is now available to be sold and shipped to anyone.

CHAPTER-5

PRODUCT SUSTAINING

The Sustaining Phase

Previous chapters focused on building and launching a new product from scratch. However, the game is just starting as a product is completed and launched. The new product is usually expected to live for several years to come, and during that time, it continues to be improved as new capabilities and features are added per changing market demand. Moreover, several unexpected issues are uncovered when the product is put in operation under different environments, and those issues will need to be fixed. These types of activities relate to *sustaining* the product, and in this chapter, we will focus on the key activities involved during the sustaining phase of the product life cycle.

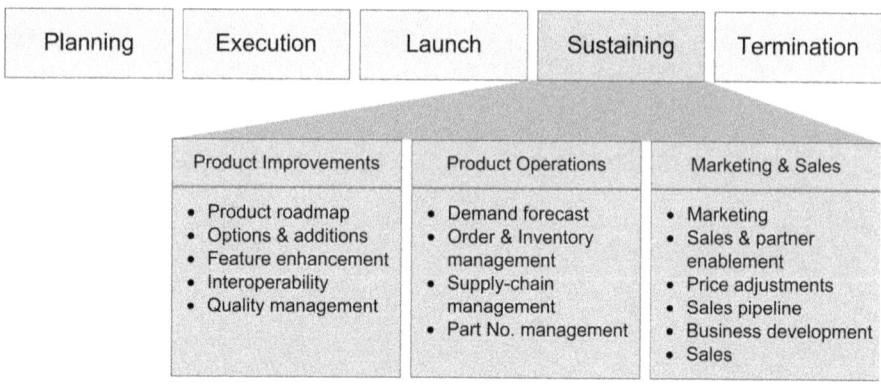

Diagram 1: The Sustaining Phase

There are several important activities related to the sustaining life of the product. One of the most important parts of sustaining is putting together an evolution and enhancement plan for the product. This includes, for example, building and maintaining a *product roadmap*. The product roadmap defines what components, features, and capabilities will be added to the product in future and how the product will be improved and expanded. In other words, how it will be sustained. Another important piece of sustaining is rounding out the functionality of the product that is necessary to keep it selling and competitive.

No product is perfect at the time it is launched. Therefore, as it is put under real-time use, potential issues will be uncovered, and those issues will need to be fixed promptly so that the customers who have already paid for the product can get the promised performance and productivity out of it. It also impacts the quality perception of the product and reputation of the company itself. Diagnosing and fixing issues is obviously an expense, and it is part of the *sustaining cost* of maintaining the product. Similarly, there are other activities involved throughout the sustaining phase, such as monitoring the sales pipeline, maintaining the demand forecast, supply chain management, and order management activities. In the following sections, we will explore these key sustaining activities in detail.

Product Improvements

Product Roadmap
Since the beginning when the product gets into the concept stage, the product manager does not only plan for the initial product version but also the long-term evolution strategy of the product. From the product strategy emerges the *product roadmap*. Which components, features, and capabilities will be added to the product in future and how the product will grow are captured in the product roadmap.

The product roadmap can be actually a collection of three pieces of information, a long-term directional roadmap, a less concrete midterm roadmap, and a short-term clear and solid roadmap. Those are referred to as the *plan of direction*, the *plan of intent*, and the *plan of record* respectively.

> *A product roadmap captures which features and improvements will be delivered on the product over a certain period of time.*

A *plan of direction* (PoD) is a very high-level roadmap that is rather vague and usually lists all the things that the product manager is thinking of delivering over a longer period of time, usually over three to five years in the networking industry. It is more of a wish list and provides the sense of direction the product is headed into without yet evaluating the feasibility of those items. A *plan of intent* (PoL) is a bit more detailed roadmap, a list of features and functionality that is intended to be implemented, but an exact delivery time frame has not yet been determined or communicated. It is therefore still an uncommitted roadmap because those items on the roadmap have not yet passed the execution commit. A *plan of record* (PoR) is a more concrete roadmap that lists specific features and capabilities, which are definitely going to be delivered over a known period of time. Those features and capabilities generally have been already prioritized, scoped, and passed the execution commit stage with resources and funding allocated. The plan of record is considered the actual roadmap most of the time because it is firm and can be shared with customers and partners.

The product roadmap is taken seriously within and outside the company. It is shared with the partners and customers only after they have signed a *nondisclosure agreement* (NDA). This is required so that none of the strategic information in the roadmap could be leaked to the competition that could plan counterstrategies, and this could be detrimental to the product. Once the roadmap has been finalized, it is kept up-to-date, and adjustments are made to it as result of periodic strategic decisions. The product manager takes the items on the roadmap through the product life cycle phases of execution and launch to productize them. We will explore product roadmap in much more detail in the later chapters.

Options and Additions

Throughout the sustaining life post launch, a product will evolve, and more options and accessories will be built around it to complete and expand its use cases and to deliver the enhancements per customer demand. Adding options and accessories rounds out the product portfolio as the product matures. In other words, new orderable stock keeping units (SKU) are added to the product family. The process of planning and introducing the options and accessories is the same as the product itself. The new components are taken through the concept and execution commits, the product requirements are written, the funding and resources are secured, the projects are created and managed, the pricing is set, and ultimately, they are made generally available. Therefore, although the product itself has entered the sustaining phase of its life, the options being added still go through the new product introduction cycle as discussed during the planning and execution phases.

One thing that is made sure of is that the options and accessories that are added by the product manager on the roadmap strengthen the product and expand its market and use cases. It is common that the sales and the partners will start requesting all kind of options and accessories as soon as the product gets out of the door. It is the product manager's job to filter out the requirements that would really make a difference. Those options should be reusable and sellable in volumes and not built for one-time sales opportunity. After the product has been in the market for few quarters, it is usually clearer to the product manager which markets and use cases the product has much better traction in based on which it can decide on the options and accessories to further strengthen the product in those applications. This is the best way to assure the product success and return on investments made.

Feature Enhancements

Just like the product will likely expand in terms of options and use cases, it will evolve and mature in terms of capabilities, features, and functionality. As discussed earlier, under the planning phase that the hardware and the software feature requests go through a prioritization process. The product manager starts gathering the new feature requests from the customers and the sales teams through a process of feature request submission. This process is only needed if a feature is being submitted by someone other than the product manager himself. The product manager also adds the new features, capabilities, and enhancements to the product himself that he thinks are important for expanding the market reach and the use cases. Not having a feature may prevent a product from being deployed in certain use cases, and hence product sales may get impacted.

As discussed in earlier chapters, the product manager reviews the submitted feature requests, filters them, and then prioritizes them based on alignment with the strategy, resource availability, business case, and other criteria that matter. The product manager bundles a set of features together and takes them through the commit process to have them developed and delivered in timely fashion. As the product matures, the type of features and functionality that is added to the product becomes narrower in focus and advance in nature. The features are less "table stakes" and more innovative. New and innovative features should be added to prolong the differentiation of the product since by then competitive products may have caught up in terms of fundamental capabilities of the product. Such features add value to the product and keeps barrier to entry high for the competition to catch up and keep the reasons for buying the product alive.

Product Interoperability

As the product gets deployed in varying customer environments and applications, it is exposed to different level of functionality, interoperability, and stress conditions. As we will explore in the next few sections, some hard to find issues could be uncovered during this process. The real production environment also provides the opportunity for the competitive products to interoperate with each other that can uncover interesting issues such as inconformity with the industry standards from one or both product sides. The products following the same standards should interoperate without issues. However, this is not always the case. This is so because standards are mostly guidelines, and it is up to the product managers on either side which part of the standards they decide to implement to be compliant or if they decide to implement in entirety. Once those issues are fixed, the product becomes increasingly stable and reliable to be deployed in complex multivendor environments.

Quality Management

Improving and maintaining the product quality at the expected level is key part of product sustaining. When we refer to the word "quality," it should be measured with reference to the customer expectations and experience. Customers set the benchmark for quality. There are several factors that contribute to the product quality. Those range from mechanical and operational robustness to the functionality and performance of the product. For example, when a product ships, all components must be able to be installed flawlessly at the customer site. This is assured through the first article inspection process discussed earlier. However, this only ensures that issues at the time of launch are handled. For example, a customer may not follow the instructions per the installation manual and may mishandle the product during installation that could cause damage to the product. It is quite possible that the customer may not take responsibility for the damage, and in business, customer is always right.

The product manager will need to find a solution and work on improving the product so that it becomes more robust, fault-tolerant, and foolproof to tolerate customer mistakes. This may require redesigning and replacing certain product components or making modifications to its packaging or documentation. Those types of activities continuing to improve the product and end customer experience are carried out throughout the sustaining phase. Majority of the issues found during a networking product's life are in the product functionality area such as the software defects. Software defects are usually easier to fix compared to the hardware defects, although the former also has different stages of difficulty and expense associated with them. As discussed under the product execution earlier, as the product progresses through the prototype, pilot, and production stages, it becomes harder and more expensive to fix an issue, especially the hardware issue. Similarly, as the product moves from the development to production stage, it becomes harder and more expensive to fix a software defect. According to some estimates, it can be *forty times* more expensive to fix a software defect or a "bug" after the product has been launched than it is to fix it during the development.

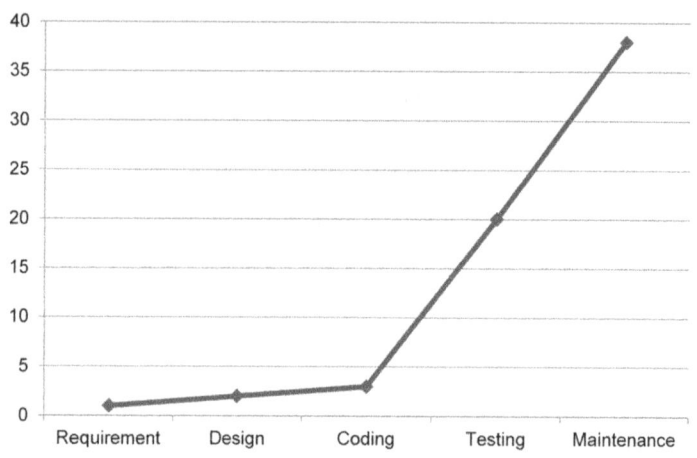

Diagram 2: Relative Cost of Software Defect Fix as Time Progresses

Generally, there is an ongoing defect triage process that focuses on prioritizing and ranking open issues and scheduling the fixes delivery for those in periodic maintenance software releases. At higher level, regardless of the company process, mostly the hardware and software defects are uncovered through (a) internal quality assurance process or (b) through customers finding and reporting them. We will discuss these two areas in more details below due to their importance as it relates to the product success.

Quality Matrix and Certifications

Many companies have a centralized quality organization to track and monitor products and services quality and customer experience. The organization uses some *key performance indicators* (KPI) to track and measure the quality. Statistics for the number of defects found, mean time to resolution (MTTR), return merchandise authorizations (RMA), and product recalls are used. The number of defects found, especially by the customers, indicates how solid the company's development and quality assurance process is. The better the quality assurance, the lesser the number of defects leaked to the customers. The mean time to resolution indicates how efficient the company is in terms of turnaround time taken to fix the defects that already have been found and reported. The shorter the mean time to resolution, the better the quality process. Similarly, the large number of return merchandise authorizations indicates the quality issues that could be rooted into the hardware or even the components and the supply chain.

The product recalls build a bad history and could be a result of several factors outlined above. A product is recalled when the issues related to it cannot be fixed in the field, and the product must be radically changed or redesigned and replaced with another product. This usually involves safety issues that are immediate danger to the end users. This could be due to several reasons such as bad design, picking up inferior quality components, manufacturing process, not complying with the regulatory safety standards, compromising on quality processes, and other reasons. Unfortunately, the product recalls do happen sometimes. However, repeated product recalls indicate flaws in company's quality processes and practices as outlines above and do hurt the brand and reputation at fast pace.

For assuring that a company follows recommended quality processes and procedures, some renowned industry certifications exist such as the *International Standards Organization 9000* (ISO 9000) and TL9000 certifications. The ISO 9000 is a set of quality management standards that ensure that a vendor company actually implements what it says in terms of its quality processes and meets the level of quality management required by the ISO certification. The ISO 9000 makes sure that the company's products meet its customers' needs while staying compliant with the required regulatory and safety standards. A different certification called the *Telecommunication Specific ISO 9000* (TL 9000) has also gained popularity in the information and communications technology (ICT) industry lately. The TL 9000 is built upon the ISO 9000 quality management system (QMS) foundations but is more telecommunications industry centric. The TL 9000 focuses on issues such as the service providers, vendors, supply chain management, contractors, and manufacturers of the hardware and software products.

In case of ISO and TL certifications, it is not a one-time event. Every year, the company is audited for its quality processes and making sure that it deserves to retain its quality certification. The product manager himself can be audited as part of the process for things such as properly documenting and archiving the requirements, product requirements documents review process and notes, requirements traceability, product roadmaps, and other such activities. The product manager can provide inputs into the quality processes and can help implement the changes to the product, but it is really the charter of the quality organization to manage the end-to-end quality process. Many customers consider the ISO and TL certifications as the quality stamp while making product purchasing decisions.

Defect Detection through Quality Assurance

A company's *quality assurance* (QA) process and its caliber have a direct impact on the product quality, its perception, and reputation, all of which affect both the business and the customer experience. The primary job of the quality assurance, a collection of multiple test teams and processes, is to uncover as many defects as possible before a product or a software release goes out in customer hands. This way, the dirty laundry stays home, and also it is more convenient and less expensive to fix those issues in time. This also helps prepare for how to message them to the customers if they cannot be fixed in expected time frame. If a customer finds a defect during the real-time operation of the product, it is not pleasant. Customers become agitated and sometimes upset; they demand the fix immediately. It hurts the product and company reputation. Sometimes it results in an escalation to the executive levels, and other projects have to be interrupted in order to provide a timely fix to the customer. This, of course, costs the company money as it pulls resources from other important priorities to fix the issues that should have been fixed already.

Some quality-conscious companies have established the rule of thumb that the number of *customer found defects* (CFD) should not be more than 5 percent of the total number of defects found. However, many companies in the hi-technology industry struggle to get closer to this benchmark. Part of the problem is that the quality assurance teams' test bed setups are generally not representative of the real customer environments, and their understanding of the product use cases may not very clear. Closer the simulation of a real environment is more deep-down issues can be uncovered before customers find them. Yet another issue is that test cases may not be challenging enough to exercise the interworking of the product. Discussing the QA process in detail is beyond the scope of this book, but these factors do matter for the product quality.

Generally, when the defects are found, they are ranked into the priority order and divided into different levels of severity. The lower the number, the higher the severity level. Therefore, a severity 1 (S1) issue is of the highest severity possible verses a severity 3 (S3) issue of moderate severity. It is more of an art than science to determine what the severity of an issue should be. The defects are tracked and adjusted in severity as the number grows. To provide the fixes for the defects found, a comprehensive *software release process* is created. A software release is a software product that is used as a vehicle to deliver new features and defect fixes to the customers. Usually, there is a hierarchy and nomenclature of software releases created to serve this multipurpose. The releases vary in feature richness, quality, and frequency. Below are some commonly used software release types in generic terms, which could vary from company to company:

- o **Major Release:** is a software release that is a superset of all features and fixes and is generally delivered from anywhere once a quarter to once a year, depending on the product and the industry.
- o **Minor Release:** is a software release that is smaller than a major release but more frequently delivered. The scope of minor releases is limited to fewer new features and more toward defect fixes.
- o **Maintenance Release:** is a software release that is issued more frequently and focused on providing timely fixes for the defects. There are no new features introduced in the maintenance releases.

When the software defects are triaged and prioritized, for every release, a must-fix list is derived. Those are the defects that are considered so important that a software release must not go out without first fixing those; hence they are also called the showstoppers. Once a release goes out to the customers, it is published with associated *release notes (RN)*—a document that lists what issues have been fixed in that release, any associated caveats or limitations as well as any important issues that are known to the customers but could not be fixed at the time.

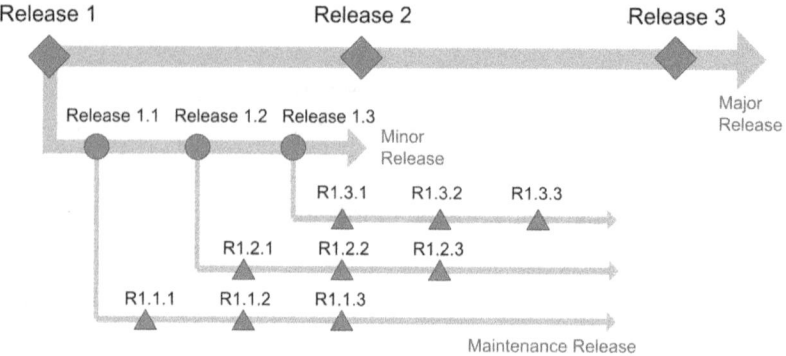

Diagram 3: Software Release Types

Defect Detection through Customers

No matter how good a company's quality assurance team and process is, some defects will always leak out. This problem can be contained but not completely eliminated. As mentioned earlier, generally speaking, if the percentage of *customer found defect* (CFD) is less than 5 percent of overall defects found, it is generally an indication an efficient quality assurance process. The seriousness of the customer found defect has been discussed earlier. When a customer finds a defect in the product, its first contact is generally the service and support team of the company, also referred to as the *technical assistance center* (TAC). The customer usually calls the technical support or files the issue through an online reporting tool based on which initial investigation takes place between a technical support expert and the customer. Relative information is gathered that will be useful to debug the issue later on to find the root cause.

From that point on, the appropriate severity level is assigned to the issue, and it is relayed to the engineering team. The engineering team assigns an engineer to troubleshoot the issue and provides a fix back to the customer through the technical support. However, before an issue can be fixed, it must be recreated or reproduced, which means that the necessary customer environmental variables are simulated as close as possible to understand how and why the problem occurs. After a root cause has been found and scope of work needed to fix it has been determined, the time required to fix the issue and the associated maintenance release time frame are communicated to the customer.

Product Operations

Demand Forecast

As we discussed in the earlier chapters, before a product goes to the manufacturing first time, the product manager needs to provide initial marketing forecast to indicate the product's projected demand. This forecast is consumed by the operations and manufacturing teams to determine how many units of product need to be manufactured initially. This initial forecast may or may not be based on any *sales guidance* and may not be very accurate. However, by the sustaining phase, the product has shipped and actual sales history has been established somewhat. Moreover, a predictable sales pipeline now exists. The forecast built and maintained during the sustaining phase is therefore should be more accurate. If the product is replacing an existing product, then that also makes the forecast much more predictable since there exists a forecast history of the older product already.

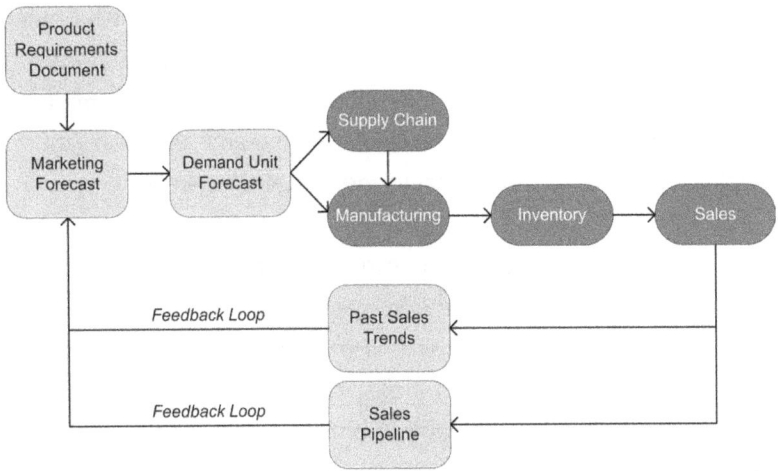

Diagram 4: Product Forecast Process

The sales pipeline is important part of the forecasting process because the product manager will need to know how many units of product should be manufactured and kept ready for a given quarter to avoid any shortage or over inventory situation. The product manager works with the operations team to have those many units manufactured and stocked. It is important to note that not all quoted products will actually translate into sales, but only a subset of it. That subset will result into actual product bookings. A booking is an actual commitment to buy by the customer and will eventually translate into revenue. At that time, revenue will be therefore recognized.

If more products are actually booked than what was indicated by the forecast, then it will result in shortage of the product and delays in shipments to the end customers, resulting in poor customer experience. This can also become a bigger issue if certain components used in manufacturing of the product have longer supply chain lead times and have to be ordered from the suppliers well in advance. On the other hand, if pipeline is large and the sales team actually books only small portions of it, the product could be over-forecasted. Lots of extra units will be manufactured and held in the inventory that was not sold, resulting into an *over-inventory* situation. The over-inventory costs company money every quarter to stock the products into the warehouses and has an impact on the income statement and the balance sheet. Usually. the accounting team needs to define an allowance for the excess, leftover, or obsolete inventory in the financial books for the company's balance sheet accounting, referred to as the *excess and obsolete* (E&O) exposure. Either situation above makes the operations team's job hard. Therefore, it is up to the product manager to analyze the sales pipeline and help translate it into a more realistic operations or manufacturing demand forecast.

The forecast is expected to go up every year if not every quarter as do the sales targets. This indicates the success of a product and health of the business. Should any surprises arise, the product manager needs to readjust the forecast and reset the expectations. The forecast considers the factors like seasonality of the business and spending habits of the customers per industry. The product manager is usually required to present the forecast to the executive team on monthly or quarterly basis and is answerable for any downward adjustments to the forecast, which can be perceived as some sort of problem with the product, the sales, or the market that needs to be fixed.

Order and Inventory Management

As discussed in earlier chapters, supply chain management is important for the product launch readiness, but it needs to be managed on an ongoing basis throughout the product life to ensure that appropriate volume of product can be manufactured and right level of inventory is in stock to fulfill the orders in timely fashion. The timeliness is worth discussing more here and so are the details of order processing, as all of this relates to customer experience and expectation.

When an order comes in, the customer expects that the product be shipped as soon as possible, as no one wants to wait after they have paid for it. For this reason, the expected *lead time* for product shipment is published ahead of time, depending on the product and the geography. The lead times could range typically anywhere from forty-eight hours for smaller products to four weeks for large and complex products. The lead time depends on the product since collecting the product from the warehouse, assembling its parts (or not), preparing the shipment information, and getting it shipped takes time. The lead time depends on the geographic location because of the individual country regulations such as customs clearance and distance from the closet shipping warehouse varies.

> *The lead time is the published expected time frame between a customer placing an order and actually receiving the product.*

The inventory management of the product starts with the accuracy of the sales pipeline and the forecast. Looking at the sales pipeline and other situational awareness, such as competitive landscape, the product manager prepares the marketing forecast, which is consumed by the operations team to derive the real demand of the product. The operations team looks at the past orders and trends to determine the true demand of the product in a given quarter, as well as accounts for the units already in the inventory, and signals to the manufacturing or contract manufacturer how many units of product need to be additionally manufactured to fulfill the expected demand of the product in a given quarter. Most of these tasks, such as the sales pipeline and the forecasting process, have been discussed in detail earlier. Once the right number of product units has been manufactured, they are shipped to multiple warehouses around the world from where the regional orders are fulfilled. The operations team also maintains a *safety stock* for emergency purposes in situations when a large unexpected order comes in and depletes the inventory.

The inventory management is an important activity that a company operation has to manage on an ongoing basis. For healthy financial performance of a company, there should be enough product units manufactured in a given quarter that are good enough to fulfill the orders and ideally fully consumed by the last day of the quarter. However, obviously it cannot be predicted with this much accuracy. In many cases, especially when a product is new and historic patterns have not been yet established, an *over-inventory* situation may occur as the product manager is usually optimistic about the product sales. Over-inventory situation is not a good situation because in the company accounting books, it is considered an expense, and stocking product units in warehouses costs the company. Therefore, the operations teams periodically monitor the inventory levels with the product manager and adjust future forecasts according to the realities on the ground.

Supply Chain Management

Throughout the sustaining phase, the product supply chain needs to be managed. As the product manager provides demand forecast every quarter, the commodity team needs to make sure that the adequate supply of the required components is secured from the suppliers. In fact, the commodity team usually secures the supply for several quarters in advance. This is to reduce the risk for any components that may be harder to obtain or may have longer lead times. Secondly, the commodity team needs to look for the alternative suppliers for cost optimization and for reducing the risk for the important product components that have single source of supply. For critical components, a dual-source approach is always a smarter approach.

Thirdly, the commodity team needs to renegotiate the costs of the existing components with the existing suppliers periodically or, if the product volumes pick up, to get the best costs possible. Negotiating the costs at the bill-of-material level periodically with existing suppliers, finding alternate sources of supply and shopping for more efficient components is the primary job of the commodity team and affects the product margins positively or negatively.

Part Number Management

As discussed during the planning phase, it is common that only certain models or part number (stock keeping units) of the same product become successful in the market and not all of them. However, regardless of the revenue, every product part number needs to be forecasted, stocked, and maintained as standalone entity in the inventory. Therefore, the overhead of maintaining all part numbers is the same. It is a common pitfall for the product managers to define too many part numbers initially and later on deal with ongoing part numbers and inventory management issues during the product sustaining life. Additionally, the product manager has to manage the part numbers database regarding allocating different part numbers for multiple stock keeping units, and it may soon become a complex problem. Throughout the product sustaining phase, the product manager and the operations team periodically review all the part numbers activity and their inventory levels as discussed earlier. The product manager reviews the sales data and makes decisions if and which product part numbers should be terminated to clean and consolidate the product portfolio.

Marketing and Sales

Marketing

Since the product has been launched, this does not mean that it will now continue to sell by itself. Even the greatest products do not sell by themselves. In order for the customers to buy a product, they first need to know that it exists, what it is about, how it can benefit them, and why they should buy it. Therefore, the importance of marketing and awareness building cannot be ignored throughout the product's life and during the sustaining phase.

Companies successful in making good business out of their products focus on and invest in the marketing significantly.

During the sustaining life of the product, it is important that new marketing campaigns are designed and launched. Those "lightening rod" campaigns can be targeted worldwide at a particular geography or a specific competitor. The campaigns must include activities that will continue to generate awareness about the product and bring attention to it, promote its differentiation and advantages, and generate new leads for sales. As the competitive landscape changes and the competition announces similar or better products, which will happen, the product manager and the marketing teams will also need to design and launch specific promotions that can offer attractive incentives for the customers and partners to motivate them to buy the product.

In addition to the above, the product collateral must be kept up-to-date, and new marketing content should be published throughout the product's life. For example, the product data sheet must be kept up-to-date as new features and functionalities are added to the product. Other customer-facing content such as solution briefs, and customer presentations must be refreshed as the product use cases expand as a result of new partnerships. Ongoing marketing attention to the product combined with an actively moving product roadmap signals to the customers and partners that the product is alive and kicking. They feel that their investments are safer. This helps customers to come back and buy more product and services resulting in repeat business. Since the product marketing is such as broad topic, we have dedicated a full chapter to this topic, and more on this will be discussed later.

Sales and Partner Enablement

One of the extremely important aspects of making a product successful is how to enable the sales teams and channel partners in making them successful at selling. Only a successful sales team and partner base can help make the products successful. Throughout the product sustaining life, the sales- and partner-enablement activities continue in terms of product promotions, collateral, messaging, positioning, competitive, and selling. We will explore this topic in more details under the product marketing chapter later on.

Sales and Partner Training

Just as the sales and partner training is required the very first time a product is launched, it is required throughout the sustaining life of the product. First of all, the human nature requires it. No one is going to master the product, the pitch, and the story the very first time it is told. Repeated training assures it until they get it. Second, as the new components and features are added to the product per the roadmap and as new markets are opened, training on those new additions and what applications they enable will be required. Most likely, the product target applications would have refined as well based on its stronger focus areas. The sales and the partners will need to be retrained per the latest product positioning and messaging. Generally, a good training is segmented in terms of the content focused at two different sets of audience. One part is focused on purely the salespeople, the other part of the training focuses on the technical resources. Furthermore, the training content should be customized for the partners.

The sales training should not be taken lightly; neither is it a one-time task. Longer learning curves are not good thing in sales. Every day a salesperson is spending in learning, it is not being spent in actual selling and hence generating the revenue. Focused, repeated, and high-quality sales training builds competent sales force that can really make the difference. The sales force and partners that are not well trained are mostly confused and come back with lots of question all the time that the product manager or the product marketing has to deal with. This can cause nuisance. On the other hand, a well-trained sales team and partner base can handle most of the challenges independently and make the desired impact.

Sales and Partner Promotions

One important aspect of driving the business success during the sustaining life of the product is through creating and offering sales- and channels-focused product promotions. For this purpose, it is important for the product manager and marketing to understand what motivates the sales and the channel partners. For example, offering additional margins for a product that the product manager wants to push through the channel could be one of the strong influencers. This can be achieved through additional discounts or by offering cash incentives such as a *sales promotion incentive fund* (SPIF) or blackened rebates. The incentive can be targeted at the channel partners, at the sales team, or both. Putting the right promotions in place is not an easy task. It takes deep understanding of the sales model, competitive landscape, demographics, and timeliness, in addition to architecting a promotion that increases volumes while protects margins.

Sales, Customer, and Partner Meetings

Other sales-enablement activities that the product manager and the marketing teams usually help with include participating in the sales and customer meetings to help the sales process, travelling to the customers and partners for product and roadmap presentations, and answering any questions on behalf of the customers on daily basis.

We have already explored the customer executive briefings of such activity. Additionally, there are sales and partner conferences every year, sometimes twice a year that require active participation of the product management and the product marketing in terms of presentations and training-related activities. Moreover, there could be closed forums such as the *customer advisory council* (CAC) and the *partner advisory council* (PAC) that periodically meet to discuss the key issues customer and partners face and to collect the feedback, which could prove extremely useful in recalibrating and aligning the company strategy and roadmaps. The direct participation of the product manager and the marketing manager in customer engagements adds credibility and helps the sales process because customers tend to believe that the product managers are honest and knowledgeable. And this perception is the reality most commonly. However, the product managers need to always balance how much time they can spend into the sales-enablement activities without compromising their core responsibilities.

Price Adjustments

We discussed pricing in detail earlier under the launch phase. An important activity that usually happens once or more than once during the product life is the necessary price adjustments. The product pricing that was appropriate at the time product was launched may no more be competitive. This is especially true when competitors announce new products that are built upon newer technology with better supply chain costs and are cheaper. This puts pressure on the sales team to make the deals at the same rate as pricing does matter in many cases and more in some geography than the others. It is therefore very important that the product manager is constantly monitoring the competitive landscape and product sales analytics and planning appropriate pricing actions if deals are increasingly lost based on pricing.

Generally, if the product pricing or discount structure is no more competitive, there will be frequent high discount requests coming from sales and partners to the product manager for approval. Those are referred to as the *nonstandard pricing* (NSP) requests. A high number of nonstandard pricing requests indicate that the product pricing is no more competitive, or the discount limit is too tight, and it is time for the product manager to recalibrate the pricing or relax the discount structure.

> *The nonstandard pricing (NSP) results when a product is sold below an expected average sale price limit or with more than set maximum discount.*

If the product forecast is declining on quarterly basis, as usually happens toward the later portion of the product life, and there are not many nonstandard pricing requests either, it may be an indication that in addition to deteriorating product differentiation in terms of the product performance, features, and functionality; pricing may also be a reason. A detailed win-loss analysis or postmortem of past few quarters of sales data can help the product manager see a clear picture. If significant number of deals were lost based on pricing, a pricing action may be overdue. This usually happens when the sales team or partners start giving up on bidding a product based on their experience that the price will not be competitive even after a nonstandard pricing approval or if they have to request too many nonstandard pricing requests. Under those circumstances, the product manager should baseline the competitive pricing again, look at the past discounts, and adjust the pricing so that it becomes more competitive. The real solution at that point however may be to start thinking of the product refresh and migration plans to a newer product.

Sales Pipeline

Throughout the product sustaining life, salespeople register the potential sales opportunities into a system or tool maintained for this purpose. If a company does not have such a tool, the opportunities could be manually documented and tracked by a *sales operations* team that provides the support services to the sales organization for day-to-day operations. When a salesperson registers an opportunity, it also assigns a probability level (such as 75 percent or 50 percent) to the opportunity indicating the likelihood or confidence level that a quote will actually translate into an actual booking.

As more and more opportunities are registered into the system, this starts building up a *sales opportunity pipeline*. The sales pipeline now looks like a long list or queue in which opportunities are relatively ranked and stacked. Every company may have variation in terms of probability levels, what they mean, and how they are assigned; but generally, those probability levels imply similar progression:

- 10% probability: *conversation is just started with the customer*
- 25% probability: *had few meetings and interest is building up*
- 50% probability: *conversation is entering into a serious phase*
- 75% probability: *short-listed and in final conversations stage for the deal*
- 90% probability: *commitment to buy has occurred*
- 100% probability: *product is booked, waiting for the purchase order*

The probability levels therefore indicate how a given "deal" is progressing. The average time it takes from the initial customer conversation to booking the order determines the typical *sales cycle* it takes for a product to sell and depends on the product and the target market.

A sales pipeline consists of all potential sales opportunities ranked in terms of different probability levels for turning into actual bookings.

The product manager may monitor the sales pipeline closely and on weekly basis to ensure that product is having good traction in the market, depending on the access to the sales data. On the product health "dashboard," this is one of the most important health meters that the product manager may want to monitor. The product manager can request from the sales operations team a weekly snapshot of the pipeline and other data desired. A healthy product should have a growing pipeline, and a significant portion of the pipeline should translate into bookings, meaning actual sales. For example, if the sales team forecasts one-third of the pipeline at 75 percent probability level and only one-fourth of that half actually translates into bookings, it must be investigated whether sales is over-forecasting the opportunities or if there are other issues due to which the opportunities were lost at such a late stage. This win-loss analysis or postmortem should be a routine process for the sale operations and the product manager to assess the competitive success of the product.

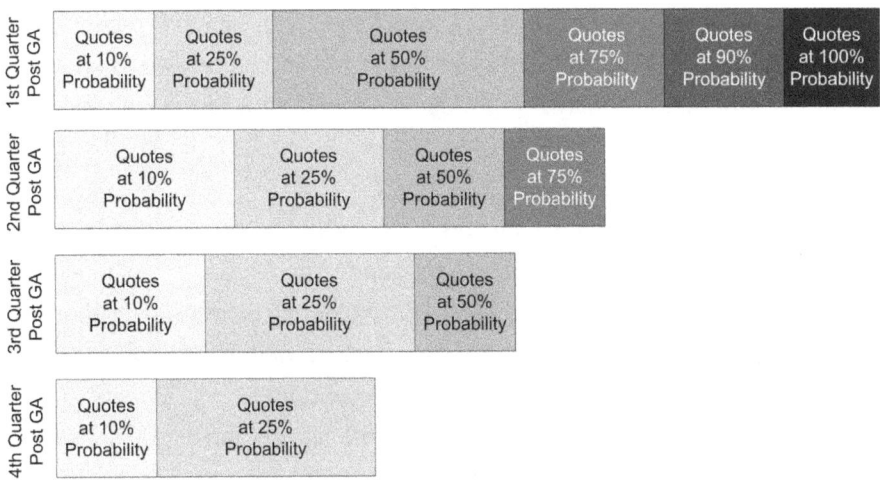

Diagram 5: Sales Pipeline Example

The sales pipeline is also important to be monitored by the head of sales team to see why different deals are *stuck* at one place for long time, as the pipeline should always be moving. A moving sales pipeline has opportunities converting to bookings at the top of the pipeline and new opportunities entering at the bottom of the pipeline. A stuck pipeline for a given opportunity may indicate that the salesperson needs to follow up with the potential customer and updated the status of the opportunity in the pipeline to move it forward or drop it and spend the sales time on other opportunities. A stuck or slow-moving pipeline overall indicates a much severe problem that could be related to the wrong market, the product-related issues, or the sales execution-related issues.

On the other hand, a "high leak" pipeline that has a poor rate of pipeline-to-bookings conversion indicates purely a sales execution problem. If opportunities get dropped out of pipeline at high probability levels (such as 50 percent and above) and if it happens too often, it requires a win-loss analysis for the lost deals and a postmortem process to find out the root causes of why those deals were lost after spending significant sales cycles on those deals that would have rather been spent somewhere else. If opportunities get dropped out of the pipeline at low probability levels (such as below 50 percent) and if it happens too often, it indicates that there is an issue with sales training and opportunity qualification process. The *qualification* implies that the salesperson has applied the provided criteria to narrow down the right market, right use case, and right target customer that would not result in waste of time.

A healthy product pipeline indicates that enough opportunities are lined up and there is good market demand for the product. It is the direct and channel sales and the business development teams' responsibility that they continue to sell to the existing customer base as well as develop new sales opportunities and build the product pipeline ongoing basis. Once there is a healthy rate of sales pipeline buildup after the launch, the product manager can start tuning the product demand forecast more accurately on an ongoing basis and start adjusting the product roadmap to align and strengthen the features that ensure its success into the areas of top sales.

Business Development and Sales

In the very outer shell of the company's business operations, the sales team and the partners continue to sell the products that the product manager has delivered to them. No matter what every other entity in the company does to ensure the success of the product and the business, nothing can be fruitful if sales team is not performing up to the mark or if the channel partners are not pushing the product. The *sales execution* is the ultimate make or break issue for the success or the failure of a product. It is said that: "A great salesperson can sell you the hell, and you will get in the line to get in."

This basically means that a good salesperson can even sell a not so good product. So if a good product is not selling as expected, it possibly points to the sales execution-related problems. The greatest companies therefore maintain greatest sales teams. They invest in them, groom them, and reward them.

A great sales team is a collection of individual personalities. Every salesperson matters. The salespeople tend to develop certain habits over period of time, which are hard to change. In a changing and highly competitive business environment, this may be the biggest obstacle in the way of a new go-to-market strategy. It is therefore important that the head of worldwide sales, which is probably the most important position in a company after the chief executive officer himself, carefully builds its sales team *top down*. When a company is trying to change, changing the sales leadership only does not help. This is a common mistake. The whole sales teams need to be rebuilt or retrained. And along with that, the whole partner base may need to be recalibrated, realigned, or reshuffled. Some of the channel partners may need to be let go while new ones may need to be added. This is an extremely important topic, and some of this is covered under the go-to-market stagey discussion in later chapter, but it is unfortunately beyond the scope of product management.

Most sales teams are *quota driven*, while most salespeople are *coin operated*. This simply means that most of the time, salespeople are tempted to sell the products that retire their quota faster. The sales leader sets the sales achievable targets for every quarter of the fiscal year based on the calculations that how much revenue must be achieved in that quarter. Those target revenue numbers are set by regions or *theaters*, where the product must be sold. Every salesperson in those regions has a pre-calculated share of the overall sales target to achieve, which sets his or her personal quota target. If everyone meets their individual targets, then overall sales targets for the given quarter will be met. This is commonly referred to as *meeting the numbers* in the sales language. If few people miss their targets and no one else in the team has exceeded their targets to compensate, overall numbers will not be met. Similarly, if some or all theaters miss their numbers and no other theater is able to compensate for the miss, overall quarterly earnings of the company will be below the projections provided to the shareholders. This will impact the company's stock price. If this happens often, the investors' and shareholders' confidence in the company starts deteriorating and stock tumbles. Bad things can start happening. Many companies end up failing despite having great products and solutions because of this single reason—their revenues do not reflect their innovation value; and at the end of the day, it is all about the financial performance of the company.

There is also another common issue with the sales process. As the product life extends, sales team tends to get into their comfort zones and turn into a collection of "farmers"—that is, they continue to use their existing customer relationships and *seed* the existing customer accounts for repeat sales to retire their quotas. This results in flat to declining company revenues with little or no growth. It is important that a significant portion of the sales team is engaged in "hunting"—that is, looking for new customers and opportunities and winning them from the competition. The *business development* (BD) team that is a subset of the sales team is primarily chartered with this task alongside the sales *account management* team. However, the business development team also looks for new ways to grow the business overall, such as finding new markets, sales channels, partnerships, and strategic alliances. The importance of the sales execution cannot be stressed enough, and in fact, a book can be written on this subject by itself. We will discuss more on sales topic under go-to-market in later chapters.

Key Takeaways

Now that the product is out there, it needs to be taken care of like a baby. Sustaining a product takes attention and effort. It is the learning time for everyone. The learning in the sense that everything the product manager and others had thought they had it covered is not really fully covered. Customers have huge potential to surprise and find missing things. This is the time to turn a good product into a great product and filling and leftover cracks. This is the time to take the product on a journey toward completeness, richness, and maturity.

Remember:

- Importance of marketing and awareness building cannot be ignored throughout the product's life.
- A product roadmap captures what features and improvements will be delivered on the product over a certain period of time.
- The quality should be measured with reference to the customer expectations and experience.
- If more products are actually booked than forecasted, it will result in delays in shipments to end customers, resulting in poor customer experience.
- If the forecast is large and the bookings are fewer, it will result in an over-inventory situation.
- The lead time is the published expected time frame between a customer placing an order and actually receiving the product.
- A sales pipeline consists of all potential sales opportunities ranked in terms of different probability for a possible booking.
- The nonstandard pricing results when a product is sold below expected average sale price or with more than set discount ceiling.
- The sales execution is the ultimate make-or-break issue for the success or the failure of a product.

CHAPTER-6

PRODUCT TERMINATION

The Termination Phase

As discussed so far throughout the book, a product's life consists of several phases, and we have by now reviewed all of them in detail except one. No product lives forever, and eventually, every product will need to be removed from the market. There are several reasons why a product does not live forever. Market conditions change over time, competitive landscape changes, newer technologies emerge that enable newer products far more cost effective and feature rich. The product that might be state of the art few years back may now become obsolete when compared to newer competitive products or to fit the evolved customer requirements.

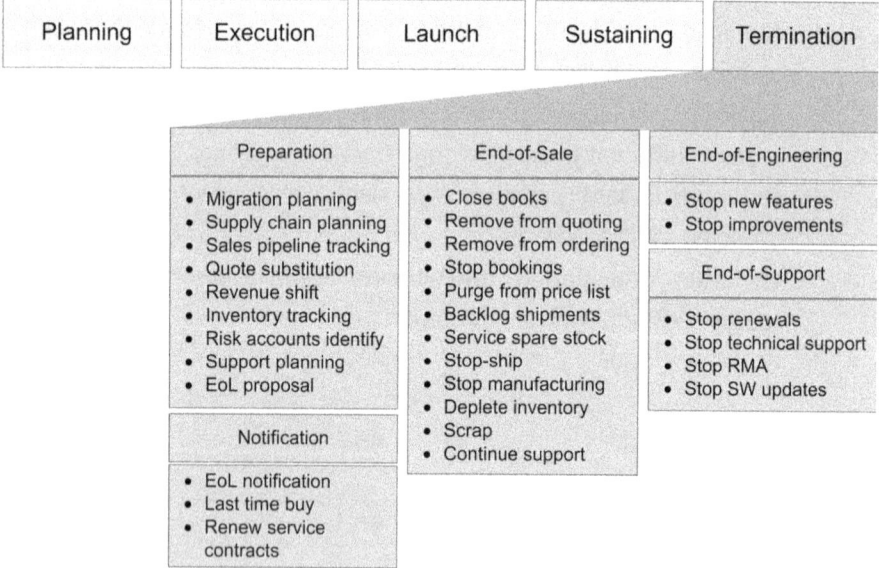

Diagram 1: The Termination Phase

It is also quite possible and is common in fact that the same vendor company might have come up with a newer replacement product to protect its customer install base and help them transition to the newer product.

Therefore, the product manager may need to plan to fully and gracefully terminate the older product. This is more often the case. As the product managers build newer products, they terminate the older ones. The product termination is not a sudden act; rather it happens in a planned way. There are several steps involved in terminating a product, more often referred to as the *end-of-life* (EoL) process of the product; and in the following sections, we will explore those steps in detail.

End-of-Life Policy

A product termination cannot take place overnight. The customers and partners must be provided with enough advance notice in order for them to prepare for the transition and have least impact on their business operations. For this reason, companies publish their product termination policy, known as the *end-of-life policy* publically. A product's end of life involves several steps. The end of life itself is a series of activities that need to be completed before terminating a product forever. Once a product has been end of life, it is no more manufactured or sold to the customers. Therefore, no maintenance, support, or fixes for the defects are provided to the customers. The end of life is an overall process that includes more than one milestone such as end of sale and end of support. Later in this chapter, we will explain those milestones.

> *The end of life (EoL) of a product is an overall process involving a series of steps involving the end of sale, the end of support, and finally the end of shipment.*

The end-of-life policy explains the order of different milestones and the time frames at which they occur, as well as what rules apply at those milestones. For example, a company may define a policy to announce the end of life of a product six months prior to actually pulling the trigger on its sales and may define what rules will apply once that has occurred. The company may further define what its policy will be regarding accepting new orders, renewing any service contracts, replacing any faulty components, and so forth during different stages of the end-of-life process. The end-of-life policy is proactively published so that there are no confusions or disputes later on, and customer can review the policy as part of making their purchase or product migration decisions. In fact, many customers do inquire about the end-of-life time frames when they purchase new products to ensure that their investments are useful over longer period of time, and their businesses will not be disrupted.

End-of-Life Process

As mentioned earlier, the product's end of life is not a single step but rather a process that consists of a series of steps that happen one after the other. Below, we will explore these steps and the important terminologies as they apply to this process.

End-of-Life Preparation

The thought process and preparations for terminating a product start long time, sometimes months or years before the actual end-of-life process is kicked off. This usually starts with a product crossing its maturity stage and getting old enough that the product manager should start thinking of how to replace it and what the replacement product should be as well as how it would be better than the older product. There is usually lots of learning done with the older product, the competitive landscape has changed, and better technologies and components are usually available. The product manager has all of that at its disposal. In addition, it has a wide customer install base that can makes product adoption convenient. The product manager prepares the concept proposal for the new product like before, and other than the new market opportunities, it also includes the migration strategy from older to newer product. It prepares the revenue analysis and what would be the impact if the replacement product was not offered timely enough. The impact could include a dip or permanent decline in revenues due to customers migrating to competitive offerings.

The product migration requires a strategy in itself. The replacement product should offer equal or better functionality, features, or performance. The existing customers must be made aware of it well in advance to help them plan the migration on their end and minimize the impact on their business operations, which also closes any cracks for competition to get in. And yet majority of the customers may expect some sort of incentive offered to them to migrate. So charging premium for additional value that the new product may offer for migration purposes and under an end-of-life situation may be pretty hard. In fact, the product manager may end up offering significant discounts. However, the company still gains since not only it has protected its existing customers, but also there is incremental revenue opportunity due to the migration and more future sales.

There are several things that need to be done before the target end-of-life time frame nears. The core team program manager or a project manager prepares a project plan to stack the end-of-life tasks and deliverables. Just like it takes lots of coordination and supervision to build and launch a product, it takes significant effort to terminate it gracefully and without a bad customer experience. Unfortunately in many companies, the end-of-life process gets minimal attention because it may be lot less exciting. Assuming that there is a project manager to manage the end-of-life process, the project manager kicks off the end-of-life process per the product manager directions. There are several groups that are impacted internally by a product's end of life and need to participate fully in the planning and execution of it, such as:

- The product manager starts working on the end-of-life proposal including a migration strategy, replacement product, end-of-sale forecast, expected time frame, etc.
- The product manager provides the replacement products mapping to the sales operations team to be entered into the quote tool and prepare the messaging to the sales teams to encourage them to start quoting the new replacement product instead of the older product. By doing so, the product manager executes a "make before break" strategy and shifts the revenue safely while bringing down the inventory levels on the older product to minimize the cost of scrap.
- The operations team starts planning the materials, end-of-sale forecast, inventory planning, spend projections, and *excess and obsolete* (E&O) exposure.
- The supply chain or commodity team starts planning the last-time-buy (LTB) components and disposition of any company assets.

- The sales operations team starts planning for any sales impacts, risk accounts, mitigation, and it starts the sales pipeline analysis and maintenance.
- The services team starts planning for the support plans, outstanding product defects as well as for the return merchandize authorization (RMA) including repair plans, and the last-time-buy forecast to the operations team.
- The engineering team starts preparing for the end-of-life impacts, hardware and software support plans, and end of engineering.
- The finance team starts planning for the excess and obsolete inventory, revenue, gross margin, and burden rate impacts. Usually, the accounting team needs to define an allowance for the excess, leftover, or obsolete inventory in the financial books for the company's balance sheet accounting.

Once the above organizational analysis is complete and necessary inputs have been complied into a proposal, it is ready to be taken for the executive approval.

End-of-Life Proposal

When a product manager thinks that it is time to terminate a product, it kicks off the process typically with an internal end-of-life proposal. Just like the concept proposal was built and presented to a review committee at the start of the product's life, an end-of-life proposal covers the case for why the product should be end of life. The proposal consists of the following sections:

- Identifying the products to be end of life and why
- Analyzing the financial impact of the end of life
- Proposed migration plan to a new product for the customers

We will discuss the details on this proposal later on in the chapter. Once the proposal is ready, it is reviewed and possibly approved by a committee for go ahead. Once approved, the product manager can issue an *end-of-life notification* to the customers and partners and start the countdown toward stopping the product sales, support, manufacturing, and shipment.

End-of-Life Notification

An end-of-life announcement is a notification to the customers and partners that the company intends to discontinue a certain product. The notification is sent to the customers and partners as well as posted online on the website where it is easily available. The notification includes the intended end-of-sale and end-of-support dates for the product. A company cannot just discontinue selling, supporting, and manufacturing a product quietly without any advance warning to the customers and partners. A typical end-of-life notification in the networking industry could be as early as six months or even one year for important customer accounts before an end-of-sale date.

The purpose of the notification is twofold. First, it alerts the customers and partners so that they can plan a transition, if needed, to an alternate product to avoid any impact on their businesses operations. Second, they can place any orders before the end-of-sale date to stock more products if needed to fulfill their foreseeable demand, typically referred to as the *last time buy* (LTB). The situation can also be in reverse. That is, the supplier may announce the end of life on one of the components that are used in manufacturing the product, and the product manager may need to suddenly execute on a last time buy to sustain its product or plan for a product refresh on its end as a result. Getting an alternate component built on demand may be very expensive and typically requires longer lead times and higher volumes or a *minimum orderable quantity* (MOQ). This also makes inventory management harder or may result in ultimate scrap. In such situations, a dual supplier strategy always helps although it is not always possible, such as in the case of a merchant silicon end of life. In case of a planned end of life, the operations team may need to place the orders with the supplier as the last time buy as well anyways.

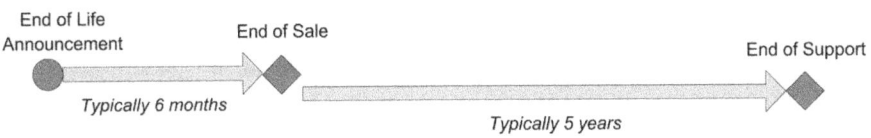

Diagram 2: End of Life Milestones

Once the notification has been issued for the end of life of a product, clock starts ticking. Customers generally still have time to place any orders for several months and even purchase new service and support contract if they did not in the past before the end-of-sale date. As the advertised end-of-sale date approaches, the product manager submits a request to the product data team for the *close of books* followed by taking it off the company's price list and asking the operations team to stop shipping the product. We will review these activities in the following sections.

End of Sale

The *end of sale* (EoS) is a milestone at which the product (and any associated services) stops selling and no more orders are accepted from the customers or partners. Even if the end-of-life announcement has been issued, there is usually plenty of time to purchase more product units and to renew any service contracts before the end-of-sale date arrives. Once the end of sale occurs, no more orders are accepted. Key items to be monitored post initial end-of-sale announcement are as follows:

- Last time supply planned and purchased (last time buy)
- Monitoring the inventory levels
- Tracking and analyzing as the end-of-sale date approaches for running out of inventory sooner than expected
- Tracking if an end-of-sale extension may be necessary
- Monitoring the sales of end of life versus alternative products
- Sales pipeline monitoring and maintenance (converting to alternatives)
- Management of any customer orders in the backlog
- Financial impacts to consider

The orders that have already been accepted may still ship to the customers. It is possible that the inventory may run out on the product before announced end-of-sale date arrives. This should be made clear in the end-of-life policy and the notification that the end-of-sale may occur by the date advertised or *before* that date in case of inventory run out, whichever happens sooner. It is up to the product manager to decide if he would like to have more inventories stocked until the end-of-sale date arrived or not. The product manager does so as a result of monitoring the sales pipeline and in case there are any significant revenue opportunities still in the pipeline. If the product manager over-forecasts at that stage, it may end up an over-inventory situation that may never get consumed because of the end-of-sale situation. Or alternatively, the end-of-sale date may need to be extended. Therefore, it has to be a balanced approach and a careful call.

Once the end of sale occurs or if the inventory runs out, the product manager sends a reminder to the sales and partners indicating a "sold out" status. At the same time, the product manager works with the order management and sales teams to analyze any outstanding customer orders and shipments in the backlog. Under a migration plan, those orders may need to be flipped to the replacement products. The product manager or the operations team alerts the pricing team that the inventory has been depleted and that the product must be removed from the price list as well as from the ordering and quoting systems so that no more orders can be quoted or booked. For the orders that were already booked, the operations team continues to work with the order management team for the last customer shipments. Once those orders have been fulfilled and the backlog is cleared, the product manager starts the process of closing the financial books or the *close-books*.

In situations where there is still excess inventory, the product manager and operations team can check with the services team if there are any last time buy needed for the service *spares stocking* such as for the product returns and replacements. If there is still inventory left, it ends up being scrapped and written off in the company's financial books. Although it is rare, the product manager may extend the end-of-sale date in order to consume the leftover inventory or for any other reasons. The end-of-sale date, therefore, is subject to change.

End of Support

The *end of support* (EoSP) is a milestone until which an already end-of-sale product is supported per the active service contracts. The services may include providing technical support, *return material authorization* (RMA) to exchange malfunctioned hardware components, and providing fixes

---

and software updates for any defects. When the end of sale occurs, from that point onward, only the product units that have already been sold and shipped are supported by the services team. Those customers, whose maintenance contracts may expire, can still renew the contracts until the end-of-support date. The hardware will continue to be replaced through the return merchandize authorization process as usual. Once the end-of-support date has arrived, all service and support activities halt, and the product is no more supported in any ways. No more returns are catered, and no more software updates are provided to the customers. At that point, the services organization also wipes its hands. Since end-of-support date comes long time after end of sale, it is expected that most customers should have migrated onto the alternate products by then. If a customer faces any issues after the end of support, the vendor company per its end-of-life policy is justified not to provide any services. Therefore, not migrating to a replacement product and continuing to operate a product until its end-of-support date is a risky proposition for a customer or business.

End of Engineering

The *end of engineering* (EoE) is the milestone applicable to software products after which no new improvements or features are provided in the form of software releases. In the networking industry for example, a typical end-of-life policy is an end-of-support date of anywhere from one to five years after the end-of-sale date for hardware products depending on the product type. For software, the end-of-support date is typically three years after the *first-customer-shipment* (FCS) date of each major release or the final release in end-of-life case. The end-of-engineering takes place with the first customer shipment of the next major release, which means that no more features or maintenance releases are provided for that major release anymore. The company may still issue patches for specific or critical issues in cases where customers have no work-around or migration path provided to the newer software.

End of Quote and Forecast

Once the end-of-life announcement has gone out internally, the product manager starts working on the end-of-life readiness. The product manager starts monitoring the sales pipeline closely and advises the sales team to start quoting the replacement products (if any) instead of the product going end of sale. By doing so, the product manager tries to shift the revenue over to the newer products, which protects the existing customer install base and shifts or even increases the product revenue and associated services revenue. Since generally the newer products promise better margins, this also results in better profitability. The process, however, should be started well in advance of the end-of-life notification. The product manager also starts forecasting lower volumes to manage the inventory accordingly. Ideally, most of the customers have been shifted to the alternate products, the forecast hugs the ground, and the inventory runs out little before the planed end-of-sale date while there are no more outstanding orders.

Close-Books and Purge from the Price List

The closing of the financial books or *close-books* is opposite of the open-books. If we remember from the launch phase, the open-books involved opening the company's financial books to start recognizing the product revenue. In the case of close-books, as the name suggests, the accounting books are closed, and no more product revenue is accounted for. The product manager submits a close-books request when the product's end of sales has arrived or the inventory is depleted, and the product manager does not wish any more orders to be booked. At which point, the product manager wants the product to be removed from the ordering system and the price list. It is worth noting that removing the product from the ordering system and removing it from the price list are two different tasks. If the product is removed from the ordering system, it is no more bookable; however, the pricing may still stay in the ordering system and on the price list, which could cause confusion. Similarly, if the product is removed from the price list only, it can still be booked since it is in the ordering system. Therefore, both should be completed. Once the close-books happens, visibility to the financial reporting is lost.

Stop-Ship

A *stop-ship* is opposite of the begin-ship. Once the close-books have happened and any outstanding orders have been shipped, the operations team stops the manufacturing process per the product manager's directive and puts the product in a stop-ship status. As discussed earlier during the sustaining phase, a temporary stop-ship can happen any time during a product's life due to different reasons, such as unavailability of the components needed for manufacturing, a quality issue, or other reasons. This time, due to end of life, the stop-ship is forever. The life of the product is now over!

End-of-Life Proposal

The end-of-life process typically starts with an end-of-life proposal. Just like the concept proposal was built and presented to a review committee at the start of the product life, an end-of-life proposal covers the case for why the product should be end of life. It consists of the following topics:

- Identifying the candidate products to be end-of-life and why
- Analyzing the financial impact of the end of life
- Proposed migration plan to a new product for the customers

Identifying the End-of-Life Candidate Products

The first step is to identify all product components that need to be end of life. It should be done carefully and thoughtfully. Once the end-of-life trigger is pulled, it cannot be easily reversed. The product manager needs to take a full inventory and financial snapshot and think of the possible impacts on the business. All possible risks must be evaluated. Before the product manager can convince others, he must convince himself that a product should be end of life. It must be considered that any of the candidate components are not shared by any another product that is not being end of life. Once the products or components have been identified, the product manager lists them at the start of the proposal.

Inventory Snapshot

After identifying the products and components that must be end of life, the product manager takes the current inventory snapshot for those items. This is very important that appropriate time for the end of life is chosen when the inventory levels are lowest. Writing off the inventory in company's financial books has direct impact on its financial performance. But for analysis purposes, the write-off amount that the inventory is worth should be quantified into the proposal. In addition, a plan for how to consume or purge that inventory by the end-of-sale date should be advised. For example, the product manager may come up with a promotion offering certain incentives for customers and partners on those items to rapidly deplete the inventory.

Sales Trend

The product manager is also required to or should include in the proposal latest sales data of the candidate products. Generally, the candidate product components should have a declining trend followed by very low quarterly volumes, so there is least impact on the business. It would not make sense to end the life of a product that is selling in reasonable volumes and still generating reasonable revenue. The low volumes could have been the result of planned or unplanned actions. For example, it may be planned such that the product manager may have launched an alternative and newer product to shift the revenues over from the older to the newer product well in advance. This helps gracefully shift the customer base over with least impact due to the end-of-life process. Or it could be planned because the company does not want to cater to the target market anymore and wants to exit it. Therefore, directing the sales force not to sell the product anymore. On the other hand, it could be unplanned due to losing the market share to competition for different reasons. The sales trend shows the rationale of why the product should or should not be end of life.

Migration Plan

In addition to the sales trend, the product manager should propose a migration plan to shift the revenues over and protect the customer base in the absence of the current product. Which alternative products and solutions will the sales and partners quote and sell in lieu of the product being end of life? What will be the impact on the net revenues and margins? What will be the impact on customer experience? Those are some of the important questions to be answered before pulling the trigger.

In addition, the product manager can propose a trade-in or other migration plan to motivate customers to migrate to the alternative product, if it has not been done already. Once the proposal is ready, it is reviewed and possibly approved by a committee for go ahead. Once approved, the product manager can issue an end-of-life notification to the customers and partners and start the internal processes as discussed earlier.

Aligning Software and Hardware End of Life

Many product managers think of the interdependency of hardware and software when launching a product but tend not to think about it after the end of life has been announced. This is a common pitfall that can cause problems. For example, if a certain hardware product or its component is supported on a certain version of software and if the duration between the end of life and end of support is shorter for the software and longer for the hardware, what will happen that there will be no software support or maintenance provided to the product while the product technically has a valid period to go before it reaches its end-of-support milestone. This will result into very unhappy customers. It is therefore important that the hardware and not the software leads the end of support, and a software release should be carried forward until all the dependent products have been terminated gracefully.

Common Issues during the End-of-Life Process

There could be several issues internal to a company when dealing with the end-of-life process end to end. Below we will discuss some of the common ones:

- Unclear ownership regarding who needs to do what and when. Generally, this is a result of having a product life cycle process in place at high level but not having detailed workflow for each stage of the life cycle process and not having a RASCI (responsible, accountable, supported, consulted, informed) model in place for ownerships. It is a project management issue and not a product management issue.
- Confusion in the sales and partner community regarding when to sell the new verses old product and the cutover timing of such. This is an issue that requires clear guidance and communication from the product manager to the sale and partner.

- Confusion on the accuracy and management of the sales pipeline as it applies to the end-of-life product and converting those opportunities to the new replacement product. Again, something that the product manager needs to work with the sales operations on.
- No mechanism to count down the inventory. Orders continue to be quoted and booked regardless if there is enough last-time-buy supply or not. This results from the poor planning and communications.
- Continuous manual monitoring and manipulation of the customer orders toward the end. The order management may need to go to the customers to have the products switched to the new product. This could be taken care of automatically through advanced configurator tools and product substitution rules.
- Uncertainty regarding how the end-of-sale announcement will be handled by the distributors with no monitoring of accounting for their inventories within the end-of-sale cycle. Frequent communications to the channel partners and proper coordination with them may reduce the impact.
- Proper way to accurately account for the expected *return for credit* (RFC) inventory from the distributors back to the vendor and timing of returns.
- The close-books helps the operations team but causes issues for the existing quotes and the discount authorizations issued to the partners. Proper timing of purge from the price list to stop new quotes and bookings is important. This could be managed by stop issuing the discount authorizations by a certain date before the end of sale.

End of Life as a Sales Opportunity

When a product's end of life is announced, it generates an opportunity for the vendor company to sell its latest and greatest product to replace the older product being terminated. This is usually the case. As a product ages out and approaches its maturity, the product manager should start thinking about how to replace it with a newer product of similar type but better, to migrate the existing customer install base. The newer product can be built using the latest technology and processes and better economies of scale hopefully, as the cost of commodities or components, manufacturing overhead, and processes might have dropped. This results into lower COGS and better margins. However, opposite is also possible—that is, the cost of certain commodities or components might have gone up. Customers also prefer to stay with the same vendor and not go through the hassle of learning about a different brand unless their past experience in terms of product quality and service experience may not have been good. It is not wise to abandon the customers and let them migrate to a competitive product.

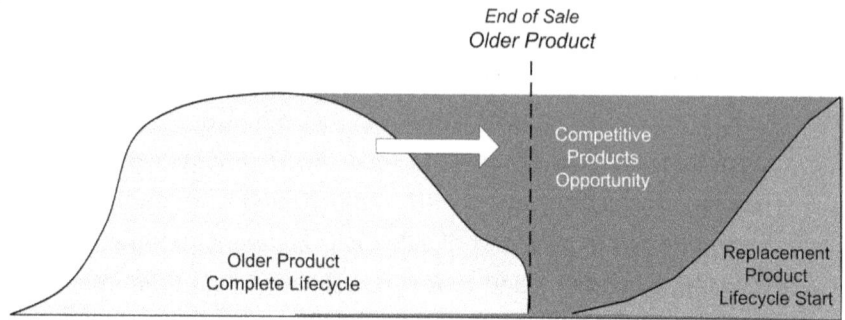

Diagram 3: Losing Install Base to Competitive Products

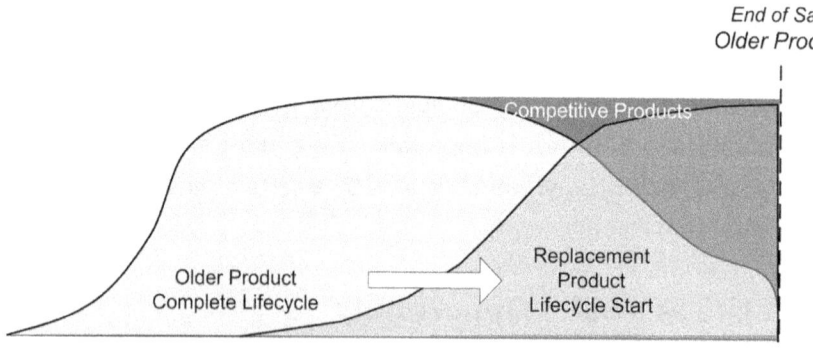

Diagram 4: Migrating Install Base to New Product

The product sales resulting from an end-of-life situation are relatively easier to close. The customers expect business continuity, and the product manager should think of providing that. For this reason, the idea of the replacement product (or product evolution) in this case should be started in advance enough so that the product can be built and launched in time, before the older product starts declining sharply. The timing is critical since leaving any gaps in there would result in creating an opportunity for competitive products to get in and start losing the market share. The product manager, working with the sales team, should keep the existing customers up-to-date on the product roadmaps and future plans. Customers want to know about those plans such that their investments are safe and their business is not disrupted. Therefore, they like to know about any end-of-life plans as early as possible and what the vendor company's plans are about the replacement solution. Existing customers can be a useful source for the feedback on existing product and lessons learned that can be translated into requirements for the next product to make it even better.

Key Takeaways

It is a sad feeling to say good-bye to a product, especially a trusted and faithful one that has generated lots of revenue for the business. However, even the greatest of the products must be given a farewell, and the end of life must happen to keep the innovation wheel rotating. New products enter the picture, and the action starts all over again. The market has either moved, or it may be a totally new market this time with new set of challenges, learning, and reward. No matter what, the product manager, sales teams, and customers always remember a great product.

Remember:

- The customers and partners must be provided with enough advance notice for the end of life of a product in order to have least impact on their businesses.
- The end of life also means later on the end of sale and the end of support.
- Writing off the inventory in company's financial books has direct impact on its financial performance.
- The product manager should propose a migration plan to shift the revenues over and protect the customer base in the absence of the current product before ending it.

- A close-books causes the accounting books to be closed for revenue recognition for a product.
- The stop-ship causes the manufacturing process to halt and stopping the shipments to customers.
- A hardware end o support should precede the software end-of-support date.
- A product's end of life may generate sales opportunities for the new alternative product.

CHAPTER-7

PRODUCT STRATEGY AND ROADMAP

Product Strategy

Building a product in fact is not a one-time event, and the product manager is not done with it as soon as it is complete. As we have been discussing, the product has a life, and it evolves and matures with time. In fact, there is only so much that can be done by the initial launch of a product. Customers will continue to ask for more features, market will continue to demand growth, and competitive landscape changes will continue to demand unplanned adjustments. The product manager is expected to plan the future of the product as far out as it can. That is how the product will evolve, what markets it will target, how will it stay competitive, and how it will fit among the overall company strategy—all this is planned into an overall product strategy. How far-out the product manager can plan things depend on the market, industry, demographics, and the type of the product. Speaking of the hi-technology industry and in particular the network industry, it is usually not of much value to plan for years ahead because technology is always changing at fast pace. Basically, a *product strategy* is an overall plan including the product evolution, marketing, market share, and profit growth targets over period of time.

> *A product strategy is an overall marketing plan that is built upon the key product features and target markets and addresses goals such as the market share and the profit growth over a period of time and how they will be achieved.*

The product strategy is built upon what is called four Ps—which are the *product* itself, the *place* that is the market and geographies it will be sold into, the *price* of the product, and finally *promoting* it through marketing and evangelism. The product strategy outlines a thirty-thousand-feet level view of how the product will evolve, how it will address the competitive challenges, what adjacent markets will be added to the go-to-market, how the product

---

pricing will be adjusted, how it will be marketed, and what sort of tactics will be used to take the market share away from competition, and so on. Goals in terms of market share, revenue, and profitability growth milestones are also identified. It is important to remember that all of the above is done under the broader vision and overall corporate strategy and should align with it. In other words, if the company's chief executive officer's vision is to grow the market share in a certain market segment, the product strategy should be designed in such a way that it is aligned with the overall vision and does not end up building market share in a an area that does not matter to the company and hence will not be sustainable. All of which would be waste of resources, time, money, and sales efforts.

EXAMPLE 1

High-Level Product Strategy

	2014	2015	2016
Product	• Launch 10G ToR • Fiber & Copper models • Stacking, 40G uplinks • SDN	• Strengthen SDN focus • Develop SDN controller • Proprietary features • Software differentiation	• Launch 40G ToR • Ultra-low latency • Stacking, 100G uplinks • FPGA for customization
Place	• Data Center market • Top-of-Rack positioning • All theaters	• Data Center market • Top-of-Rack positioning • All theaters	• Data Center market • HPC market • HFT market • All theaters
Price	• $200 per port ASP	• $150 per port ASP • Extra margins through software	• $1500 per port ASP • Extra margins through software and purpose built hardware/software
Promo	• Channel sales • Heavy direct touch • Price/differentiation • Incentives for partners	• Channel sales • Heavy direct touch • Solution partners • Differentiation selling • Incentives for partners	• Channel sales • Heavy direct touch • Differentiation selling • Incentives for partners

---

The product strategy is not a static plan, but it is a living and evolving document. The product strategy can be as simple as a PowerPoint slide, it could be a detailed document, or it could be just whiteboard discussions and meetings. As a common practice which may vary from the company to company, key executives form a *strategy panel* and meet periodically, such as once a month, to make the strategy decisions. The strategy panel may include the head of product management, head of engineering, head of marketing (CMO), chief technology officer (CTO), and the chief executive officer (CEO) among others. The strategy discussions involve discussions like what directional changes company needs, how to grow the top line and bottom line, what sort of solutions and products are needed to accomplish that, what competition is doing, what adjustments need to be made to the existing product portfolio, etc. The product management team is the key participant in strategy discussions and educates the executives about ground-level realities and challenges that feed into the strategy. It is an open discussion, and although the product management has lots of input into the strategy, it is mostly the top executives who call the final shots; and whether the product managers wholeheartedly agree or not with the direction, they have to implement the things according to it.

Product Roadmap

From the strategy emerges the product *roadmap*. Which components, features, and capabilities will be added to the product in future and how the product will shape up are captured in the product roadmap. The roadmap shows the road that leads to the product future and highlights all the attractions located along that road. The product roadmap can actually be a collection of three pieces of information, a long-term directional roadmap, a less concrete midterm roadmap, and a short-term clear and solid roadmap. Those are referred to as the *plan of direction*, the *plan of intent*, and the *plan of record* respectively and are discussed below.

> *A product roadmap captures which features and improvements will be delivered on the product over a certain period of time.*

Plan of Direction

A *plan of direction* (PoD) is a high-level roadmap that is rather vague and usually lists all the things that the product manager intends to deliver over a longer period of time, usually over three to five years in the networking industry. It is more of a refined wish list and provides sense of direction the product is headed into without spelling out too many details. The plan of direction focuses on the macro-level items rather than micro-level features. The plan of direction is the firsthand expansion of the product strategy into an executable plan of action. No dates or time frames are usually provided. The plan of direction is used mostly for internal planning and not disclosed to the customers or at least every customer. Sometimes it may be shared with very important customers and partners who might be making large investment decisions about those products. Not all companies build or maintain or at least document a plan of direction.

Plan of Intent

A *plan of intent* (PoL) is a rather detailed roadmap and the list of features and functionality that is intended to be implemented over a certain period of time, but an exact delivery time frame has not yet been determined or communicated. It is therefore still an uncommitted roadmap and not yet processed through the execution commit process. The plan of intent starts lining up different features and capabilities together that actually can be turned into a plan of record and executed upon as every item on the plan of intent goes through the process of achieving commitment, funding, and implementation. The main difference between a plan of intent and a plan of record is that the plan of intent consists of the features and functionality list, some of which may or may not actually be implemented; there are no commitments made and no exact delivery dates provided. It is just intent so it is subject to change without any strings attached.

Plan of Record

A *plan of record* (PoR) is a more concrete roadmap that is specific and lists all the features and capabilities that are definitely going to be delivered over a certain period of time. Those features and capabilities generally have been already prioritized, scoped, and passed the execution commit stage of the product life cycle process with resources and funding allocated. Therefore, there is least risk in communicating those plans to the customers. We will explore the feature request and prioritization process in the later sections in this chapter. The plan of record is considered the actual roadmap most of the time because it is firm and can be shared with customers and partners.

A basic plan of record consists of a time line, usually within the next three years, and different product features lined up along the time line with each of them having a delivery time frame. The plan of record is more granular within near future, and it becomes less granular in terms of time frame for the items farther out. For example, as a rule of thumb, it may be possible to list delivery time frame within week accuracy if a feature is coming up in the current fiscal quarter. If a feature delivery is within the next couple of quarters, the time frame could be listed within month accuracy. If it is farther out, only the year could be listed. This "reverse funnel" scheme creates some wiggle room and provides some flexibility as items in distance future are harder to predict with accuracy and in case any unexpected delays do occur in delivering a feature.

The plan of record is an expansion of both the plan of direction and the plan of intent into more detailed and a committed plan. It is in fact a plan of intent taken forward with firm implementation plan and dates of delivery specified. It is called the plan of record because it gets is "on the record" with customers. Unlike the plan of intent, the plan of record is a "promise to deliver" listed items and it is not easy to back out or make changes to it once it has been communicated. The sales teams will make deals based on the plan of record commitments. Even the delay in delivery time frames may result in consequences if certain commitments were made to the customers. Therefore, every plan of record usually begins with a legal disclaimer. A delay in the delivery of a feature may result in financial accounting challenges as well such as *revenue recognition* (RevRec) issues, which is discussed later.

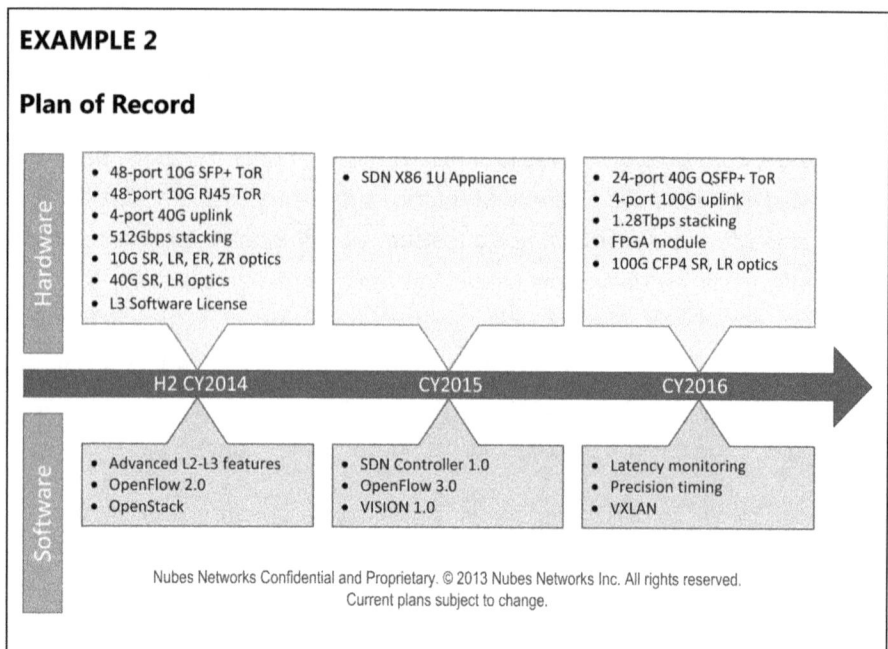

EXAMPLE 2

Plan of Record

Hardware
- 48-port 10G SFP+ ToR
- 48-port 10G RJ45 ToR
- 4-port 40G uplink
- 512Gbps stacking
- 10G SR, LR, ER, ZR optics
- 40G SR, LR optics
- L3 Software License

- SDN X86 1U Appliance

- 24-port 40G QSFP+ ToR
- 4-port 100G uplink
- 1.28Tbps stacking
- FPGA module
- 100G CFP4 SR, LR optics

H2 CY2014 — CY2015 — CY2016

Software
- Advanced L2-L3 features
- OpenFlow 2.0
- OpenStack

- SDN Controller 1.0
- OpenFlow 3.0
- VISION 1.0

- Latency monitoring
- Precision timing
- VXLAN

No matter which type of the product roadmap, it is a serious piece of information, and it is shared with the partners or customers only after they have signed a *nondisclosure agreement* (NDA). The agreement is required so that none of the strategic information could be leaked to competition or mishandled in any other way. The competition could plan counter strategies that could be detrimental for the product's and the company's business. This is discussed later in detail.

A plan of direction (PoD) is a high-level, less-detailed roadmap of uncommitted features that are of strategic importance and define the direction in which product will evolve.

A plan of intent (PoL) is a detailed roadmap without any commitments or exact delivery time frames, subject to change.

A plan of intent (PoR) is a detailed and firm roadmap that consists of committed features and their delivery dates.

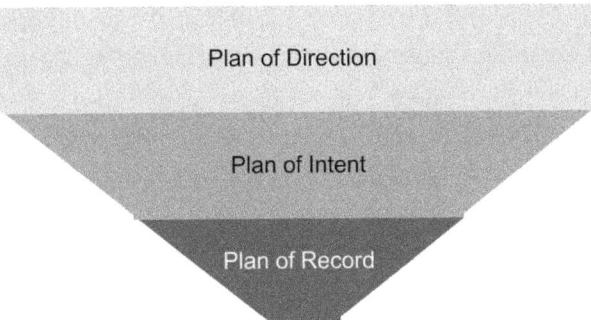

Diagram 1: Product Roadmap Funnel

Roadmap Prioritization

As discussed earlier under the planning phase, the hardware and software features go through a prioritization process before they find a place on the plan of intent and eventually on the plan of record. In case of the software productizing, that is usually in the form of new software releases, the product manager starts gathering the new feature requests from customers and sales teams through a process of *feature request* (FR). This process is only needed if a feature is being submitted by someone other than the product manager himself. Feature request process can apply to both hardware and software products although it is rare for a hardware idea submission. A feature request outlines the overall business opportunity, potential customer name(s), description of the feature requested, and the time frame by which it is requested. There is usually a feature request form or template that the sales representative fills and submits to the product manager.

The product manager collects all feature requests submitted in a given time frame window, reviews them, and *accepts* or *rejects* the requests based on alignment with the company strategy, resource availability, business case, and other criteria. Out of all accepted feature requests, the product manager then ranks them in terms of relative priority based on business impact and urgency to market. It is usually hard to judge the direct revenue impact for software features and prioritize them purely based on that basis, unless it an independently sellable software product. In case of a software product supporting a hardware product, indirect revenue contribution, and enablement of sales for the hardware product, as well as the strategic impact of the feature should be considered and so should be the *lost opportunity cost* due to not having the feature. Not having a feature may prevent a product from being deployed in certain use cases, and hence product sales may get impacted. The feature prioritization process has been discussed in detail in the earlier chapters.

Once a feature has been accepted, the product manager lists it under an intended time frame on the plan of intent. As next step, the product manager bundles a set of features together into the product requirements document and takes it through the concept and execution commit stages per the process discussed earlier. Those features which pass the execution commit are now eligible to be listed on the plan of record and can be shared with the customers and partners under the nondisclosure agreement.

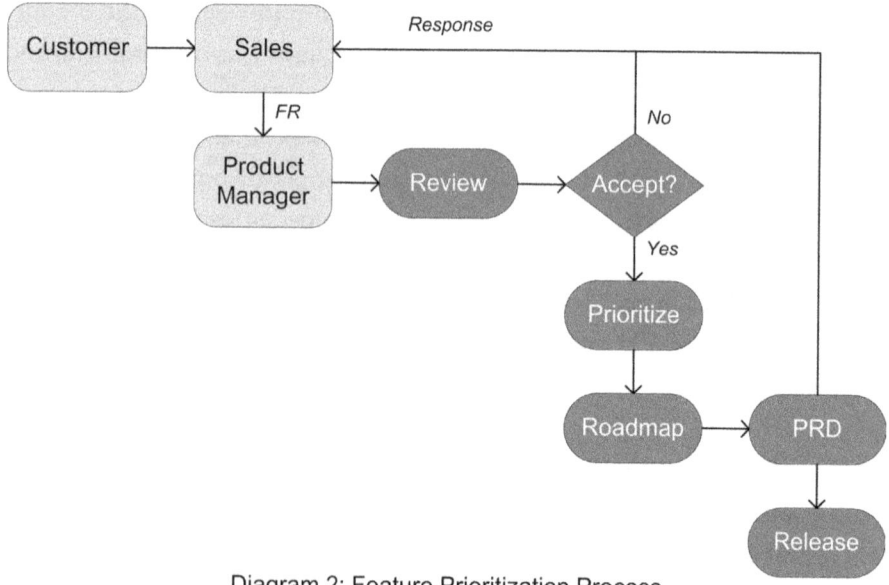

Diagram 2: Feature Prioritization Process

Legal Considerations

Generally speaking, any time a statement is made to the customers, especially in written, regarding intent or delivery of certain products or features in future, the customers tend to have it interpreted as a promise of delivery. In other words, they hear what they want to hear. A product manager therefore needs to be extremely careful what to communicate and what not to communicate to the customers as well as when to communicate and how to communicate. If a customer *assumes* things based on certain statements made by the product manager or someone else on behalf of the company and if the customer's business is impacted later on because of that miscommunication or misunderstanding, the customer could take legal action against the company—whether reasonable or unreasonable.

Legal Disclaimers

A plan of intent or plan of record is probably the most important document a product manager can publish, and it has to be protected by the appropriate *legal disclaimers*. A legal disclaimer is meant to alert the customers and exclude the company from any legal liabilities as a result of any misunderstandings. Example of a typical legal disclaimer is shown below that could be displayed at the start of a product roadmap:

> *This product roadmap represents the Nubes Networks current product direction. All product releases will be on a when-and-if-available basis. Actual feature development and timing of releases will be at the sole discretion of the company. Not all features are supported on all platforms. Presentation of the product roadmap does not create a commitment by Nubes Networks to deliver a specific feature. Contents of this roadmap are subject to change without notice.*

In addition to the legal disclaimer, the footnotes of the roadmap document also have legal statements displayed such as below:

> *INTERNAL OR NDA USE ONLY*
> *Nubes Networks Confidential and Proprietary. ©2013 Nubes Networks, Inc. All rights reserved. Information contained herein presents our current product direction and is subject to change.*

As it is clear from the legal disclaimers and the footnotes, whether the roadmap is being presented in person or an electronic or printed copy of it is being provided to a potential customer or partner, it provides the safety barrier required for any future changes without being on the hook and later on can be used as a legal defense if needed.

Nondisclosure Agreement

A *nondisclosure agreement* (NDA), as the name sounds, is an agreement that someone signs agreeing that he or she will not disclose the details being discussed or knowledge being obtained regarding a matter to another party. In other words, the nondisclosure agreement ensures the confidentiality of the important matters being discussed. A product roadmap is a strategic document and important matter. Leaking any details of it to the competition can severely hurt the business of the company. The competition could learn about the future strategy and can plan counter tactics or could deliver the product capabilities and features even sooner than the original product. This could fail or impair the product differentiation in the marketplace and cripple the business. Therefore, it is very important to understand that the product roadmap is not to be shared with just anyone and everyone. It is a "need to know" based document.

It is, however, required sometimes by the customers and partners who plan to make significant investments into buying the product, that the product roadmap be shared with them. The roadmap is important to them since they want to ensure the future proofing and investment protection knowing that the product they are purchasing and putting into their businesses will not go away in near future. Also that the vendor company has plans to evolve the product, and the customer will continue to receive new features and functionalities as well as the required level of service and support. For this reason, usually a nondisclosure agreement is signed with the customers to assure that they will not share the details with anyone else. For such reasons as well, the product roadmaps have legal warnings as below:

> *The sales team is prohibited from using this roadmap version in customer discussions openly. This NDA roadmap version*

can only be presented to customers by the product manager. Customer must be under NDA.

Commitment Letters

Sometimes, customers are not satisfied by the plan of record with legal disclaimers. They want a solid commitment that a certain feature will be delivered precisely by a certain date. This may be critical to the customer for the reason that the customer may be planning certain important business activity on their end that may be counting on the new feature rollout. Therefore, the customer may ask for a commitment of some sort. For this purpose, the product manager or the executives may need to issue a legal commitment letter to the customer. The letter lists the required features or capabilities to be delivered by certain date, and it is signed by one of the executives. Once the letter has been issued, company is now liable to deliver the committed items on time. The commitment letter now overrides any legal disclaimers in the plan of record. Generally, a commitment letter is only issued for a feature in the near term.

Accounting Considerations

Every company has a *fiscal year* (FY) to operate on. The fiscal year is the financial year over which company yearly financial performance is accounted for and published. The fiscal year may or may not be aligned with the typical calendar year (CY) that starts on January first and ends on December thirty-first. For example, a typical fiscal year extends from July first and ends on June thirtieth. Furthermore, a company's financial performance is tracked, and its revenue is recognized on *quarterly* basis because it is much more manageable.

A fiscal quarter is basically three months fiscal period. In a fiscal year, there are four quarters. Therefore, the fiscal quarters may not be aligned with the calendar quarters. The fiscal year may be defined according to several important factors such as customer's budget spending patterns, seasonality of the business, and other demographic or geopolitical background. It provides a company much better control on its financial performance based on the nature of its business.

When a product roadmap lists the delivery of certain features per certain time frames of a fiscal (or calendar) year, and if that feature has any revenue associated with it, it then must be delivered within that time frame to appropriately recognize the company revenue per the accounting rules if a customer has made a booking for it. This is also true if a commitment letter has been issued to the customer promising the delivery of certain features by certain dates. For example, if the delivery of a product or a feature was promised to be completed and delivered by the first quarter (Q1) of fiscal year 2014 (1QFY14) based on which a booking has been made, it is now obligatory on the company to ship the product to the customer within the expected time frame to the customer. If for some reason the product gets delayed, the revenue for the booking cannot be recognized and included in the financial earnings for that quarter, and a *revenue recognition* (RevRec) issue has occurred.

	Jan-Mar	Apr-Jun	Jun-Sep	Oct-Dec	Jan-Mar	Apr-Jun
Calendar Quarters	Q1 CY2013	Q2 CY2013	Q3 CY2013	Q4 CY2013	Q1 CY2014	Q2 CY2014
Fiscal Quarters			Q1 FY2014	Q2 FY2014	Q3 FY2014	Q4 FY2014

Diagram 3: Fiscal versus Calendar Year Example

The accounting department has to note down the order and account for it appropriately in the books as *deferred revenue* for the next quarter or the quarter according to the slip in the schedule. The principle of revenue recognition comes from the fundamentals of accounting principles based on matching the transactions, that is, revenues and expenses in a given period. According to the fundamental principle, revenue can be recognized when a product or a service has been transferred regardless of when the actual money is received. According to the general rule of accounting, revenue cannot be recognized until the following two conditions have been met:

- The product or the service promised has been delivered, and cash has been received for it. Revenue now can be realized.
- The product or the service has been transferred, and delivery is completed.

Otherwise, it is considered a *deferred income*, which is a liability in the financial books.

Key Takeaways

The fact is that when the product manager thinks of a new product idea the very first time, it is the time to think about the holistic picture and vision of how the idea will evolve over time. Thinking big does help when it comes to long-term strategy. Not all product managers are necessarily good strategic thinkers. Strategic thinking is a separate gift that comes from the abilities of seeing the big picture, connecting the dots, and possessing analytical skills. It is a test of using past experiences and current knowledge to predict the future. Now that the product has been launched, the product manager has more traction on developing interesting things around it and enjoying the game of strategy.

Remember:

- The product strategy should be aligned with the overall vision and strategy of the company.
- The product strategy is an overall marketing plan that is built upon the key product features, targets, and goals.
- A product roadmap captures what features and improvements will be delivered on the product over a certain period of time.
- Both the plan of intent and the plan of record are important pieces of information that should be shared on need-to-know basis.
- The product delivery time frames on the plan of record should be communicated such that there is enough room for any shifts.
- The sensitive and strategic information should always be protected through legal disclaimers and nondisclosure agreements.
- The commitment letters override any legal disclaimers.
- If a product or feature commitment has been made and a booking has been made based on the commitment, it must be delivered on time or a revenue recognition issue will occur.

CHAPTER-8

GO-TO-MARKET ROUTES

Go-to-Market Strategy

Until now, we have discussed in great detail how to plan, build, and deliver a product from scratch. In this and in the next chapter, we will study how to make it successful once we have the product. No matter how great a product or solution is, without solid marketing plan and execution, it may turn out to be a failure. If the product is considered to be the "water," then the *go-to-market* (GTM) is the "pipe." Just like different diameters of the pipes, a storage and distribution mechanism delivers the water to the homes; a go-to-market strategy plans for what mechanisms and channels to use and how to extend the product reach to the right markets and customer concentrations where it can be sold. An average product with well-executed go-to-market strategy has better chance of success than a great product with poor go-to-market execution.

So what is really the go-to-market strategy? The go-to-market strategy is part of the overall *marketing strategy*. Whereas the overall marketing strategy determines *what* products and services a company will offer and *who* it will offer those to, the go-to-market strategy deals with *how* that will actually happen. Basically, a go-to-market plan is the overall strategy on how the customers will be found and contacted; what *channels* will be used to reach them; what partnerships, solutions, processes, and tools may be used to develop the business; and how the product's value proposition will be taken to the customers. Furthermore, how the company will invest into certain areas to expand that value over time.

> *A go-to-market (GTM) plan is the overall strategy on how the customers will be found and contacted; what channels will be used to reach them; what partnerships, solutions, processes, and tools may be used to develop business; and how the product's value and differentiation will be taken to the customers.*

There are several components of the go-to-market strategy, but the most important one is choosing what *market routes* will be used to take the product to market. Companies build different strategies around the products, such as how much heavy lifting they can do and how much brand leverage they have got. But before that, the top-level decision that the company executives need to make is that which sales model to use to go to the market—*direct sales* or *channel sales*—or maybe a combination of both. Next, we will explore more details about those go-to-market sales models.

Direct Sales Model

The *direct sales* is the most challenging sales model for any company to establish and maintain, but it has its reasons and advantages. The direct sales is a model in which the company hires and maintains its own sales team to reach out and interact with the end customers directly without any intermediaries involved. The direct sales model makes sense when a company:

- has strong focus on certain market verticals usually with high barrier to entry.
- does not yet have strong brand and reputation for articulating differentiation and value proposition to the end customers.
- already has market reach in some cases and now seeking bottom line growth by keeping all product and service profits to itself.
- has target customers who are well informed, trained, and ask deep questions before making a purchase decision.
- involves opportunities and use cases of the product that are complex in nature, requiring a highly skilled and trained sales force.
- wants to offer its own services offerings on top of the product.

> The direct sales is a model in which a company hires and maintains the sales team to reach out and interact with end customers directly without any intermediaries. involved.

The major advantages of the direct sales model are as follows:

- A company can keep offering a range of products and value-added services to improve business as it desires. Complete control on the offering portfolio.
- High-value, high-touch representation of the company, which elevates its brand.
- Highly skilled sales force that puts a much better impression on the customers and actually fights to get the deal all the way compared to the channel model.
- Invest and train the sales force once, get long-lasting returns.
- Once the customer relationships are established, a company pretty much owns those relationships and customer contacts as its valuable asset.
- Opportunity to enjoy maximum products and services margins.
- No conflict of interest compared to channel sales where a channel could also represent and sell competitive products.

There are also shortcomings with this model such as:

- It is an expensive sales model since a sufficiently large sales force needs to be hired, trained, and maintained worldwide. Usually, it is only justified for large companies with multimillion- or billion-dollar revenues or companies having a narrow-focused premium market.
- It may be an over investment when launching a new product or if the company is new to the industry when compared to a channel sales model.
- The initial ramp-up time is longer to train and develop the sales force expertise up to the required skill level.

Channel Sales Model

The *channel sales* is a model in which a product vendor company leverages other companies, their sales resources, and market reach to sell its products. Those other companies become the *channels* or *routes* to the market for its products and services. The channel sales model is relatively less heavy when compared to the direct sales model, and it also offers greater expandability and flexibility. The companies that offer this service become the *channel partners* of the vendor company supplying the products. The channel sales model makes sense when a company:

- is trying to enter a new market and does not want to invest in building and maintaining a direct sales team.
- has strong brand name, which sells by itself and does not rely on heavy differentiation or value proposition-based selling.
- wants to leverage other companies established market reach and relationships for rapid revenue growth.
- wants to reach to multiple geographies and markets faster and wants to give an appearance of a larger company to its customers.
- is seeking top-line revenue growth and can compromise on profits.
- product does not require complex discussions and is easier enough to be sold by a third party with some training.

> *The channel sales is a model in which a company leverages other companies, their sales resources, and market reach to sell its products instead of maintaining its own sales team.*

The major advantages of channel sales model are:

- *Plug-n-play* or *setup-n-go* approach. Much faster initial flow of revenue compared to the direct sales model, provided that right channels are selected.
- Tapping into key customer accounts by leveraging third-party relationships that otherwise may not have been easier or quicker for a company to establish directly.
- Global reach and expandability if, when, and where needed by adding new channel partners to the portfolio.
- More word of mouth.
- Opportunity to augment services with the partner to expand the range and reach of service offerings.
- Appearance of a large company sales force with broader market reaches.

There are also some shortcomings with this model such as:

- The channel partner co-decides which products and services subset it can handle and offer to its customers.
- Sharing the products or services profits with the channel partners.
- Constantly enabling and training the channel partners in terms of product, solutions, and messaging, etc.
- Sometimes not having the end customer visibility or contact.
- The customer relationships and leads belong to the channel partner and not to the company.
- May not be an everlasting model unless the product volumes are significant through the channels.
- May cause conflict of interest when a channel partner also represents and sells competitive offerings.
- Is usually margin or reward driven—that is, the channel partners push the product on which they make most markup or overall rewarded.
- Requires the supervision and management. Can cause channel conflict among multiple partners trying to sell the same product to the same customer and may require a sophisticated deal registration process.

The products that have less competition are very appealing to the channel partners since this translates into easier sales and better margins for them. They prefer large manufacturers with best of the breed solutions and companies with niche products. They also prefer to develop solutions they can offer under their own brand along with service offerings. It is worth discussing that there are multiple types of channels and channel partners in the networking industry, each offering different types of services and with different types of business models. We will explore those channel partner types next.

Resell Channel Partners
The *reseller* is a type of channel partner who takes the vendor company products and uses its own market reach to sell those products, hence making profit out of them. How much profit a reseller can make is usually bound

by the contract between the reseller and the vendor company, the supplier in this case. Usually, it is negotiated as a percentage of the product price. The reseller takes the products as they are and does not modify or *rebrand* them. It puts the products on its own price list or catalogue but sells under the original company's brand. Reselling is therefore generally a high-volume, low-touch play. That is, the reseller or the vendor either has to do little or no customization to the product. The reseller has to sell the product in large quantities to make business sense of it for both companies. Large-scale resellers maintain their own warehouses and distribution network. Their competitive advantage is generally based on quick delivery lead times, anywhere, anytime. Therefore, they generally bulk order and stock the products in advance from the vendors. For this reason, resellers provide a demand forecast periodically to the vendors that the product manager, which can be incorporated into the overall forecast that the product manager provides for the manufacturing.

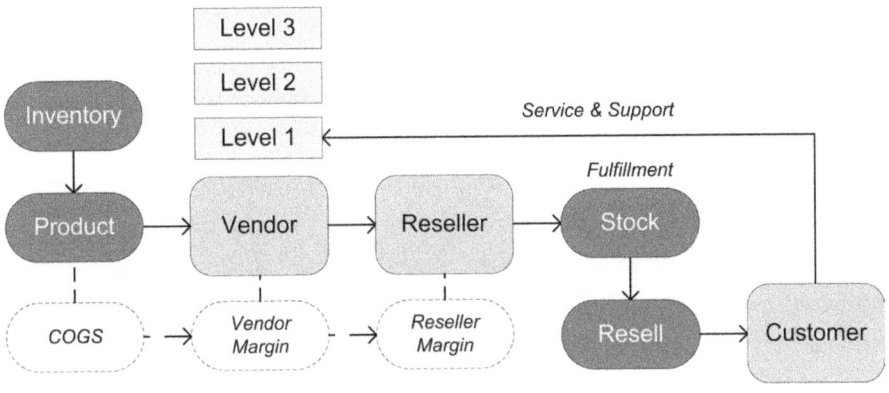

Diagram 1: Resell Channel Model

In case of reselling, generally the service and support is passed on to the product vendor as if it was selling directly, but that may not always be the case. This may include the *level-1* (initial), *level-2* (advanced level), and *level-3* (expert level) support. The service and support terms and how the end customer issues will be handled through the resell partner are agreed upon in the resell contract. The resellers' strengths are generally the variety and the choice of products and solutions they can offer from multiple sources, their broad reach with strong worldwide distribution and order fulfillment capabilities, and their sales mechanisms and customer reach such as online or catalogue. Examples of some well-known large network resellers are Dell, CDW, and IBM. These types of resellers have global reach and can land major sales opportunities; therefore, they are very selective in who they chose to become partner with and which products they carry. However, there are many other smaller resellers. The resell is an easier to set up and easier to terminate relationship without lots of strings attached.

Value-Added Resell Channel Partners

The *value-added reseller* (VAR) are the type of reseller that specialize in certain focused areas, usually much narrower than the large commercial resellers, and offer value-added services on top of plain product reselling. Generally, the value-added resellers are smaller companies that are localized and more focused on niche markets. However, there are also large-scale, value-added resellers that have multinational presence.

A value-added reseller typically takes the products and provides services such as basic rack and stack, configuration and testing, and then ships the solution to the end customer for deployment. The real value add comes from the on-top services, such as professional services in which case the value-added reseller provides consulting services with the products as well as any software applications or tools to provision a customized solution per the end user requirement. The professional services can include services such as end-of-network design, consulting, and post-sales training.

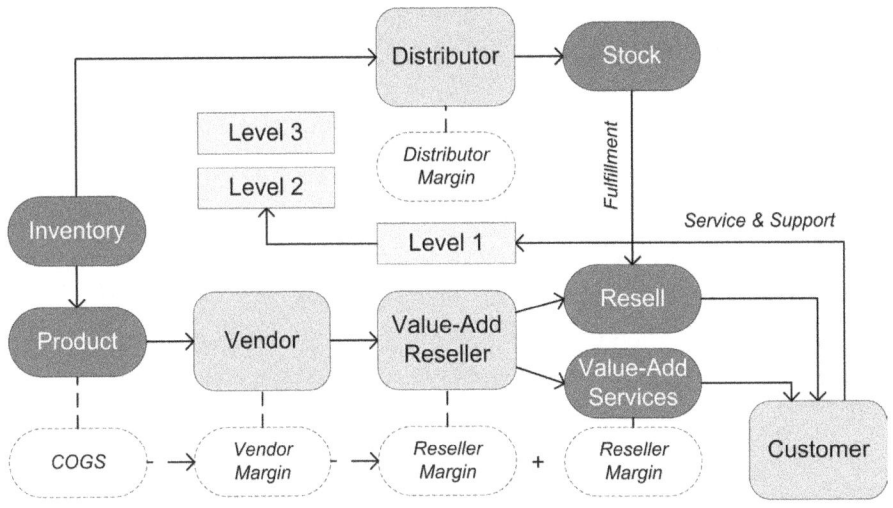

Diagram 2: Value-Add Resell Channel Model

Unlike the resellers, who bulk-order and stock the products in advance in their own warehouses, the value-added resellers generally do not maintain any warehouses and work with local *distributors* or the vendor itself to stock the products and fulfill the orders. We will explore this fulfillment model later. The value-added reseller channels are generally good way to bring the business up in remote smaller geographies especially where local relationships are valued, such as rural areas and small towns. The average deal size through the value-added resellers is smaller, the volumes are much lower and inconsistent compared to large global resellers.

System Integrator Channel Partners

The *system integrator* (SI) is a company that provides expertise for putting multiple products and components together into a complete system or solution form for the end customer. The system integrator makes sure that the integrated system functions as a whole. The system integrators maintain highly skilled talent that can provide customized solutions and tools, both hardware and software. Those solutions can be highly customized and built on demand; therefore, the system integrators can charge premium for those services. The system integrators make significant revenue from the consulting services, while rest from the design, integration, and commissioning. The system integrators are different from consultants in that they also deal with the products and procurement.

The system integrators are very popular for their services in large and complex deployments where stacks are high. Unlike the value-added resellers, the projects system integrators work on are time consuming and long term. The system integrators look for mature and stable, as well as feature-rich products and solutions that make their job easier. The system integrators are good channel when trying to get into large and complex mission critical deployments.

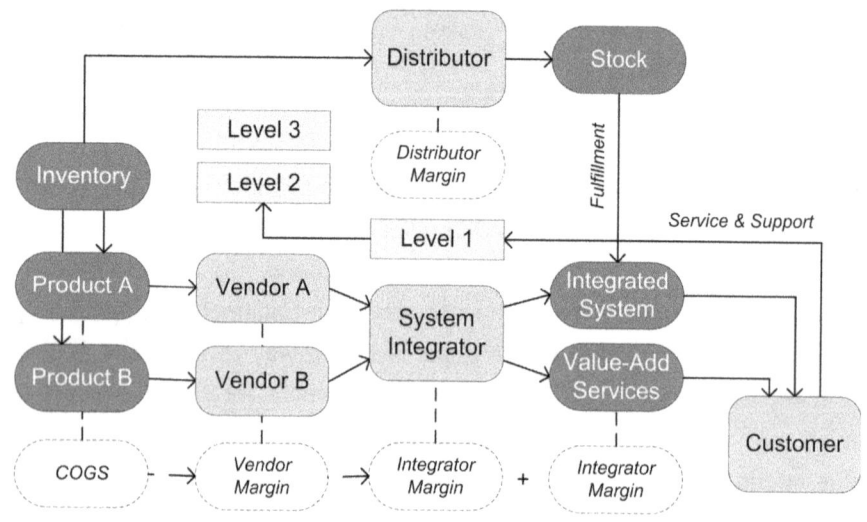

Diagram 3: System Integrator Channel Model

OEM Channel Partners

There are situations where a resell partner or another vendor company would like to take someone else's product and *rebrand* it so that to the end customer, it appears to be that company's own product without much clue about the original manufacturer. This sort of channel is called *original equipment manufacturer* (OEM) channel. The original vendor or manufacturer of the product has an *OEM-out* business whereas the company taking the products and reselling them has an *OEM-in* business. By rebranding here means customizing or converting the product in such a way that the OEM supplier vendor's brand is suppressed, and the reseller company's brand is brought to the surface.

Rebranding can range from *light* to *heavy*. For example, the light rebranding could be limited to just painting the product in different color or putting the reseller's logo on the front panel. On the other hand, the heavy rebranding may involve full industrial design changes including repainting, changing the logo, adding some sort of bezels, putting customized labels on the product, different packaging, creating custom part numbers, as well as the software rebranding including displaying the reseller's name and other information in the software displays. The product documentation and manuals may also be required to be rebranded. In addition, the OEM partner may want to offer its brand of services with the product. The idea is to remove the traces of the OEM supplier's identity from anywhere. For those reasons, the OEM, especially an OEM-heavy, can be a fairly expensive and complicated model to set up. The customization requires significant changes to a company's processes and tools and requires resources, all of which adds to the product costs on top of its COGS.

There are several reasons why someone would like to have an OEM versus resell go-to-market model. The OEM-in is generally useful for very large companies with significant brand perception who have a product or solution gap they need to fill quickly, but would not like to dilute their brand perception by exposing to end customers that they are reselling someone else's products, especially someone smaller or not so reputable. Moreover, they want to provide a seamless high-quality experience to their customers with consistent look and feel of the solution and their own services wrapped around the solution. Offering high-end branded services is one of the reasons for the OEM. If an OEM reseller has strong value-added services offering model, it would generally want to offer its own brand of services bundled with the OEM products. Services is a high margin business, and therefore, OEM resellers want to leverage their own services infrastructure in place, and they can do so much comfortably when the product appears to be single vendor based to the end customer. The reason for an OEM-in is also the price control. Since the products are rebranded and sold under new part numbers, it is a clever way to avoid channel conflict with the supplier's other channel partners. This gives the OEM reseller total control on pricing the products and the on-top services at its discretion. In this regard, in terms of control and flexibility in pricing, positioning, and go-to-market, the OEM may be the best alternative to otherwise building or acquiring a product.

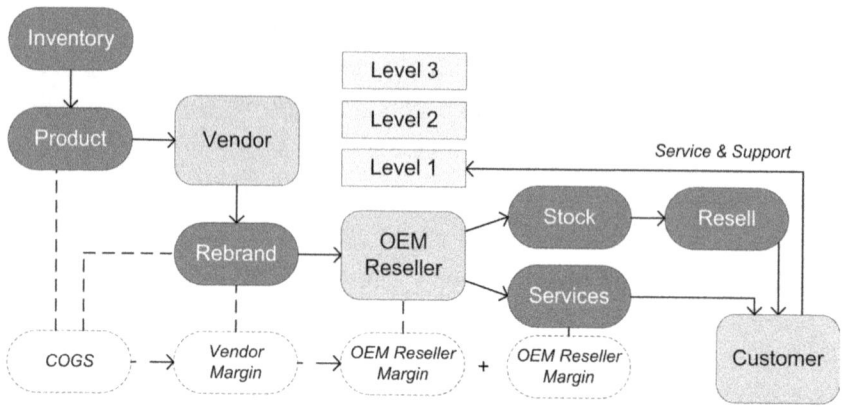

Diagram 4: Original Equipment Manufacturer Channel Model

The OEM model to begin with between two partners may be an ambitions and daunting task. While the relationship and channels are yet untested, investing too much upfront in terms of money, time, and resources in OEM activities may not prove that fruitful as one may perceive. This is especially true for the supplier, as many OEM-out contracts push most of the heavy lifting on the OEM-out vendor rather than the reseller. The OEM also makes sense when a vendor is supplying OEM-out to several resellers and not just one. The OEM as a follow-up stage of a resell relationship however may make better business sense. In such cases, both companies have usually tested the business grounds and go-to-market together, channels have been tuned up, and a run rate business with predictable volumes can be forecasted. So when the sales pipeline has been established, it may be time for shifting to the OEM model if required to take it to the next level.

Distribution Channel Partners
While most of the channel partner types discussed above involves dealing with the end customers who actually use the product, a distribution

channel partner or a *distributor* (*disti, for short*), on the other hand, does not deal with the end customers. Rather it acts as a middleman by taking a company's products, stocking them in advance, and distributing them to the resellers and value-added resellers. The distributors maintain an efficient distribution network involving their own warehouses, inventory management system, order fulfillment system, and logistics. The distributors therefore can unburden a vendor from maintaining its extensive warehouses and distribution network. The distributors order the products in bulk per their own forecasted demand and stock them in their warehouses from where they supply the products to resellers on demand. In doing so, the distributors also tackle the country-specific, customs-related regulations. Since the distributors stock the products regionally, it helps greatly in reducing the delivery lead times to the end customers and improving the customer experience.

There are other advantages that the distributors offer as well. The distributors help recruit the resellers and expand sales channels for a vendor company. In addition, the distributors help shield the product vendor from the credit risk by offering their own financing terms to the resellers. In case of bankruptcy or default of the borrower, this unburdens the product vendor from dealing with legal and financial accounting issues.

In summary, the distributors offer many advantages including the logistics, lead time, and credit offering. The multiple-tier distribution model is discussed later in the chapter.

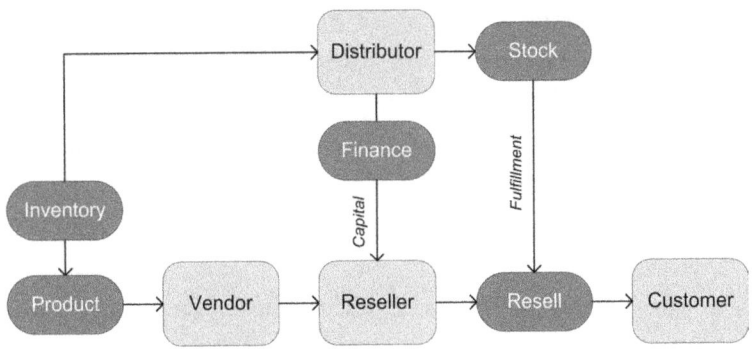

Diagram 5: Distribution Channel Model

Direct Touch Sales Model

Now that we have discussed the direct and channel sales models in detail and we can compare them side by side, it may be interesting to think of why not use the best of the both worlds. In fact, such a sales model does exist. This hybrid model is called the *direct touch sales* model. In this case, a vendor heavily relies on the channel sales using a typical distribution model; however, the vendor also maintains its own sales force, only much smaller than the one in direct sales model. The sales force in this case assists the channel partners in sales as well as hunts new opportunities along with them or by themselves. This parallel approach works well such that the company invests in training both its own sales force and the channel partner while this helps in maintaining the direct interaction or direct touch with the end customers. The sales team can engage in hunting and progressing the sales opportunities along with the channel partners and assists them in the end-to-end sales process.

With this model when the bookings are made, the orders are usually fulfilled using the distributors. The overall sales effort is shared by the sales team and the channel partners. In this way, the sales effort spent on a given deal is less than that in case of the full-blown direct sales and the sales force can handle more deals in a given time frame. The direct touch model makes sense when a company:

- has strong focus on certain market verticals usually with high barrier to entry.
- has a modest brand name but trying to elevate it.
- wants to leverage the channel partner's established market reach and relationships for the revenue growth.
- wants to reach to multiple geographies and markets faster and wants to give an appearance of a larger company to its customers.
- has target customers who are well informed, trained, and ask deep questions before making a purchase decision.
- involves opportunities and use cases of the product that are complex in nature, which requires a skilled and trained sales force.
- wants to be flexible in offering a choice of its own services or the partner offered services with the products.

The direct touch is a hybrid model in which a company uses channel sales for market reach but also maintains its own sales team to assist the channels and to reach out and interact with the end customers.

The major advantages of the direct touch model are:

- High-value, high-touch representation of the company that elevates its brand.
- Highly skilled sales force that puts a much better impression on the customers.
- Invest and train the sales force once, get long-lasting returns.
- Build end customer relationships.
- Supervision of any conflict of interest compared to pure channel sales.
- Global reach and expandability if, when, and where needed by adding new channel partners to the portfolio.
- Tapping into key customer accounts by leveraging third-party relationships.
- More word of mouth.
- Opportunity to augment services with the partner to expand the range and reach of service offerings.
- Appearance of a large company sales force with broader market reaches.

There are also shortcomings with this model such as:

- Channel partner co-decides which products and services subset it can handle and offer to its customers.
- It is still little bit expensive since a sales force needs to be hired, trained, and maintained worldwide, although of smaller size.
- The initial ramp-up time is longer to train and develop the sales force expertise up to the required skill level.
- Sharing the product and service profits with the channel partners if that is the case.
- May cause some conflict of interest when a channel partner also represents and sells competitive offerings.

- Requires supervision and management. Can cause channel conflict among multiple partners trying to sell the same product to the same customer and may require a sophisticated deal registration process.
- Constantly enabling and training new channel partners in terms of product, solutions, messaging, etc.
- Is usually margin or reward driven—that is, the channel partners push the product on which they make most markup or overall rewarded.

Fulfillment Models

Now that we have discussed a range of channel partnerships, let us review typical channel fulfillment models commonly used in case of direct touch or channel sales.

Two-Tier Channel Fulfillment Model

A two-tier distribution and fulfillment model is the one in which a vendor sells its products through two layers of channel partners. This *stacked* model includes a *distributor* as the first tier, followed by a reseller or value-added reseller as the second tier. The distributor orders the products based on its projection of the market demand and stocks them in its own warehouses. The value-added reseller hunts the new sales opportunities with or without direct touch sales assistance and "brings the deals." Obviously, the channel partners of the competition are trying to do the same. If the deal is won and when it is time to ship the products to the end customer, order is fulfilled from the associated local distributor's inventory with whom the reseller is working with.

Since the distributors bulk-order and stock the products in advance from the vendors, it is not known at the time of the shipment to the distributors who the products will be eventually sold to and when. For this reason, some vendors choose not to recognize the revenue at the time they ship to the distributors. In that case, when the product is actually booked by an end customer, the distributor sends the *point of sale* (PoS) data to the vendor for recording the financial transaction information on the end customer whom the products were sold to. At which point, the vendor recognizes the revenue according to the amount of the order. Yet some vendors may choose to recognize the revenue as soon as the distributor orders them. In this case, the distributor is treated as any other customer, and orders are booked and shipped. This simplifies the accounting process; however, no point-of-sale data is provided, and therefore, the vendor has usually no visibility into who the end customers are.

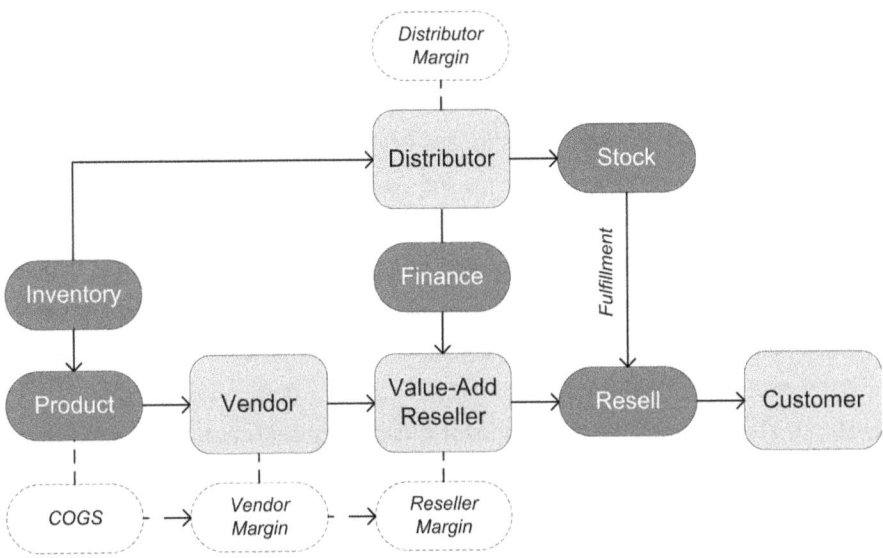

Diagram 6: Two-Tier Distribution Model

A reseller in this model may make higher margin than the distributor because of its key role in hunting the sales opportunities. Since the distributor margin is usually relatively a smaller portion, it is very much margin driven, and every percentage point matters. How much margins the distributor and the reseller make by default at maximum is bound by the legal contract. However, sometimes the margins are deal driven. Sometimes there are specific discounts granted to the channel partners per the nonstandard pricing for an end-user deal through a preapproved *discount authorization* (DA). There is a unique discount authorization number associated with every such deal in the quote tool. This gives the channel partners freedom to apply the discount when and where they need to. Later on, if a channel partner such as a reseller or an end customer references it on their *purchase order* (PO), the vendor can then reference the discount authorization number to determine what discount should be applied to the order. The distributors report this to the vendor through the point-of-sale data when they sell a product out of their inventory and the vendor gives them a credit for the amount of the discount. The mechanism of discount authorization is also used for the channel-focused promotions, as discussed later, in which case more than one end user or distributor can use the same discount authorization number.

The two-tier model is a popular model for channel sales since in this case the distributors act as if they were the company's own distribution network, reducing the need to build warehouses and stock inventory worldwide. The vendor may still need to maintain few regional warehouses to stock the inventory post manufacturing and to cut down the shipment lead times. As discussed earlier, the distributors also help recruit the resellers and expand sales channels. In addition, the distributors help shield the vendors from credit risks by isolating the financing to the resellers on their end. The limitations of this model are that a vendor has limited control and influence on the end customer. It is, therefore, important for business growth that vendor exercises direct touch effectively and tries to educate and influence the end customers by partnering with the value-added resellers. Activities such as arranging seminars and marketing events where resellers and their customers are invited are effective techniques.

Single-Tier Channel Fulfillment Model

A single-tier fulfillment model is rare, but it does exist and mostly with large-scale companies. When a company is large enough that it has its own vast distribution network worldwide, it may not need to rely on distributors and dilute its margins by sharing them with more tiers. In that case, it makes sense to recruit and deal with the resellers directly. This fulfillment model works well in conjunction with the direct sales model. The vendor is much closer to the end customers and can leverage its direct sales to influence the end customers as well as to train the resellers while it fulfills the orders through its own warehouses and vast distribution network.

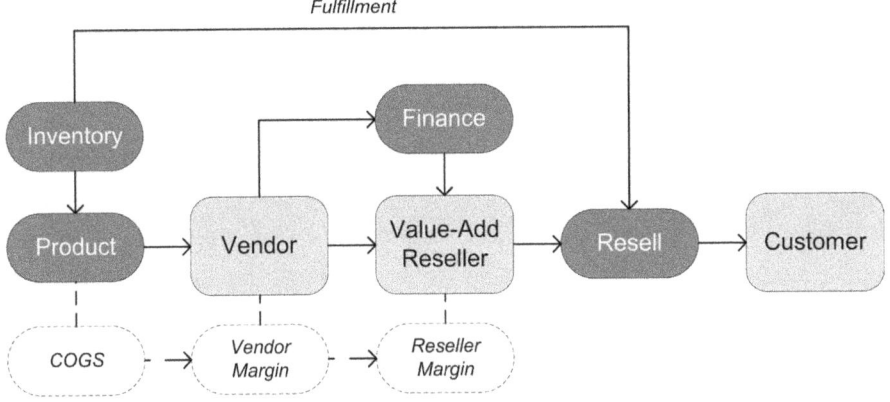

Diagram 7: Single-Tier Distribution Model

The disadvantage of this model is that the vendor has to recruit the resellers itself to expand the sales channels. Vendor also has to deal with the inventory management and logistics to stock and distribute the products around the world. This also involves dealing with customs, country-specific regulations, and staying within delivery lead times. The vendor also has to put the financing plans in place and offer credit to resellers to enable them purchase the products to sell. By doing so, the vendor takes credit risks involved as in any lending business.

Solution and Strategic Partners

Besides channel partners who are directly involved in selling the products, there are other types of partnerships as well that play critical role in the overall go-to-market strategy. Those partners may not sell products necessarily for each other, but they join hands in complementing solutions, filling any gaps in their portfolios, and strengthening their businesses together. Those partnerships can be based on joint go-to-market strategies and can use common channel partner base for extending the market reach. Those partnership types are discussed here.

Solution Partners

When two companies have products or components that complement to form a more productive solution, they become solution partners. The result is an *ecosystem* of the hardware and software product and services that can solve unique and larger problems. This fits the criteria that the total is bigger than the sum of its components. The overall *solution* that gets architected together can offer significant value. However, there are other advantages as well. In this case, two or more vendors may also leverage their brands, channel partners, market reach, and existing customer install base to generate meaningful business together with win-win results. This is an integrated go-to-market opportunity. This is also an effective strategy where smaller companies can collectively compete against larger incumbents.

The solution partnership is usually a result of *build versus buy versus partner* analysis to bridge some sort of gap in a company's offerings. Once two partners have come up with a solution together, it must be qualified and tested so that there are no issues when the customers deploy the solution. The testing is important because products come from different vendors and may never have been interoperated before. The found defects must be fixed, and any product improvements should be made by the product managers on both sides to fit the solution. Once the solution has been qualified, it is captured as a *reference design* (RD) or a *reference architecture* (RA). The reference architecture illustrates how the solution is architected, its key components, technologies, recommended environmental conditions, initial configurations, and other parameters under which the solution has been qualified. The customers, sales, and channel partners all refer to the reference architectures as the key piece to assemble and sell solutions.

> *A reference architecture illustrates how the solution is architected using multiple products and technologies, and it provides the associated configurations and other parameters under which the solution has been qualified.*

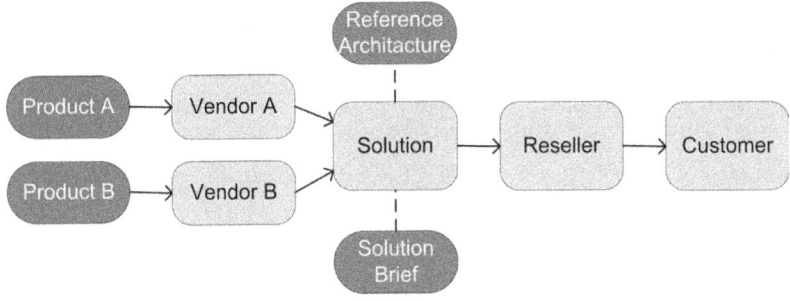

Diagram 8: Solution Partnership Model

Another important piece of material that comes out of joint solutions is a *solution brief*. As discussed earlier during the product launch, a solution brief is usually a one or two pages document on a particular focused solution. It explains the key requirements of that solution and how the selected products from partners fit those under the overall solution. While the reference architecture is too detailed and technical document intended to be used during the implementation stage of a solution, a solution brief being a marketing document is the firsthand introduction and overview of the solution. Solution briefs are popular with customers as they are short, and they provide the insight into more relative real applications for them.

Strategic Partners

The strategic partnerships are usually formed between the two companies when they have common strategic goals and agendas. This is therefore also referred to as the *strategic alliances* (SA). Strategic alliances are formed when two companies have their sight on more long term and much larger objectives than just selling the products together. In other words, the solution partnership may be a result of the strategic alliance between two companies, but the strategic alliance is usually not a result of a solution partnership. While the solution partnership can commonly be formed between two uneven-sized companies, the strategic partnerships are usually formed between two even-sized companies or companies of similar stature.

The strategic alliance, just like a war alliance, goes beyond just selling each other's products. The alliance is usually formed between the top executive tier of the companies, and they share strategic information with each other to capitalize on much bigger and long-term objectives and winning together. The strategic alliance therefore translates into cooperation on several fronts in both companies including engineering for the joint innovation, marketing for the joint go-to-market, and sales for the joint business development.

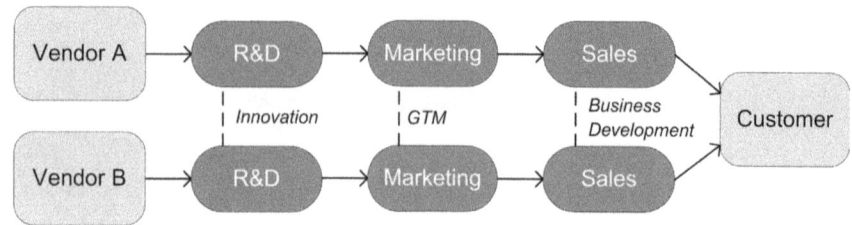

Diagram 9: Strategic Partnership Model

Deal Management Process

One of the very common challenges when selling through the channel partners is that sometimes more than one partner try to chase the same customer and compete over the same opportunity. This causes a *channel conflict*. This is a good problem to have that more than one channel partners trying to push the same vendor's products and maximizes the probability of winning a deal; however, it has to be done fairly and with least chaos. This basically means that in such a situation, it has to be decided on first-come-first-served basis. For this reason, the vendor company puts in place what is called a *deal-registration* or *deals-desk* process, which is a best practice and a must-have process.

When a channel partner has found a potential sales opportunity, it must register it with the vendor's deals desk immediately to avoid the situation in which another channel partner may come up and claim the same opportunity. Once the deal has been registered, the channel partner has exclusive right to bid for the opportunity while other channel partners wait in the queue. This way, the channel conflict is avoided. It also avoids the bad customer experience such as multiple parties trying to sell exact same product to the end customer, which may annoy the customer.

> *The channel conflict is a situation in which more than one channel partners try to compete over same sales opportunity, trying to sell the same set of products to the same customer.*

Service Offering through Channels

Services are a high-margin business, quite often the channel partners may want to define and offer their own services wrapped around the vendor's product. In some cases, the partner services are augmented with the vendor services. Based on this, there are three broad services offering models through the channel partners, a *pass-through* model, a *detached* model, and a *combined* model. With the pass-through model, a partner only resells products and passes the service offering responsibility to the product vendor company. This allows the partner to focus on its core business while maintaining the direct customer relationships. With the detached model, a partner offers its own brand of services and does not rely on the vendor company to provide any services. The partner maintains the direct customer relationships. With the combined model, a combination of the above is used where the partner offers some services, while relies on the vendor company for other services. In this way, the partner augments its services offering with those of the vendor's.

In case of the reseller channel partners, generally the service and support is a pass-through to the original vendor as if it was selling directly, but that may not always be the case. A value-added reseller typically takes the products and provides services such as basic rack and stack, configuration and testing, and then ships the solution to the end customer for deployment. The value-added service comes from the on-top or premium or professional services. The system integrators maintain highly skilled talent that can provide customized solutions and tools, both hardware and software, augmented with the advanced services. Those solutions can be highly customized and built on demand. Therefore, the system integrators can charge premium for those services. In case of the OEM partners, offering high-end and high-margin branded services is one of the attractions for the OEM relationship. We will explore more on the services offering later in the book.

Channel Support and Incentives

Driving business through channel sales model successfully requires a full channel support and reward process and infrastructure in place. Channel partners rely heavily on the product vendors for knowledge and skill transfer, training, and marketing support. Usually, vendors allocate a budget to sponsor partner marketing activities out of overall *marketing development fund* (MDF). The marketing development fund can have direct influence on how much a partner is motivated to promote one vendor's products versus another. The partner uses the funded dollars to organize business development and marketing activities such as customer events, meetings, and entertainment activities to generate the business. Even if partners use their relationships and market reach and arrange marketing events and activities, they mostly seek the support from the vendors to provide tailored marketing collateral and public speakers who are subject matter experts. We will explore more on the partner focused marketing under the product marketing chapter later in the book.

Another important aspect of driving the sales success through channels is to offer promotions for the channel partners. It is important for the product manager and the marketing to understand what motivates the distributors and resellers. For example, offering additional margins for a product that the product manager wants to push through the channel could be one of the strong influencers. This can be achieved through additional discount authorizations or by offering cash incentives such as a *sales promotion incentive fund* (SPIF) or back-end rebates. The sales incentive consists of offering cash bonuses to sales and partners for selling certain products per the required quantities or rules. The sales incentive can be targeted at the channel partners, at the sales team, or both. Putting the right promotions in place is not an easy task. It takes deep understanding of channel partner model, competitive landscape, demographics, and timeliness, in addition to architecting a promotion that increases volumes while protects margins.

While the promotions are sometimes based on offering upfront incentives such as more upfront margin or sales incentive, they can also be based on the back-end incentives such as rebates. A *back-end rebate* reimburses a percentage of the sales price, as agreed upon per contract, back to the partner *after* the revenue has been recognized for the sale. The difference between offering an upfront margin versus a back-end rebate is that the rebate is conditional on selling the product first. It puts the burden on the partner to sell more products in order to make incremental dollars. While the accounting for upfront discounts is simpler, in case of rebates, it is more complex. The sales data from the channel partners has to be correctly reported by the end of each fiscal quarter, and appropriate dollars need to be allocated to issue rebates in the form of what is called *credit checks* or *credit memos* back to the partners.

> A back-end rebate reimburses a pre-agreed upon amount back to the purchasing party out of the purchase price, after a product has been sold and other agreed-upon conditions have been met.

The partners are generally ranked into multiple tiers per a defined hierarchy to qualify for what types of service and support leverage and other incentives they can be entitled to by the vendor company. For example, the partners could be ranked as *bronze, silver,* or *gold* partners, with gold having the privileges to most incentives. Usually a point-scoring mechanism is provided to partners to earn reward points based on different criteria and motivates them to progress up through the hierarchy. Discussing the channel incentives and reward system is beyond the scope of this chapter and the book, but it gives the reader some idea.

Build versus Buy versus Partner

It is not uncommon that when a company finds gaps in its products or services offering and concludes that it is important to fill a gap either because there is significant customer demand or because there is significant business opportunity for such a product or solution. Under such situation, the product manager and executives will need to make certain decisions based on the urgency to market, whether the missing product or technology must be developed, or whether to acquire such a product or technology from outside, or whether to partner with another vendor who has it and can readily supply it. This "build versus buy versus partner" analysis is part of devising a good go-to-market strategy. Generally, following are the key factors in deciding among the three choices:

- **Building** a product from scratch is considered when:

  o Confidence in a product idea, market, and business case is high.
  o Company has the required resources and funding to build the product.
  o Company is familiar with the target market.
  o Company wants a strong hold on the product revenues and profitability.
  o Company wants to establish or strengthen its brand identity in the target market.
  o There is not a rush to market.

- **Buying** a product, such as an OEM-in or an acquisition, is considered when:

  o Confidence in a product idea, market, and business case is high.
  o Company has funding but not the time or the required resources to build the product.
  o There is urgency to the market.
  o Company is familiar with the target market.
  o Company wants strong hold on the product revenues and profitability.
  o Company wants to establish or strengthen its brand identity in the target market.

- **Partnering** for a product, such as a solution partnership, is considered when:

  o Confidence in a product idea, market, and business case is experimental.
  o Company does not have funding, time, or the required resources to build the product.

- o Company does not have the motivation to build or acquire the product.
- o There is urgency to the market.
- o Company may or may not be very familiar with the target market.
- o Company does not need strong hold on the product revenues and profitability, at least initially.
- o Company wants to establish or strengthen its brand identity in the target market.

Based on what the market demand and strategy requires, appropriate go-to-market solution can be applied. Sometimes, it is preferred to partner for a product or a solution offering first to test the grounds before making a more involved investment decisions such as buying or building it. This is, in fact, a common practice.

Key Takeaways

Defining and implementing a go-to-market plan for the product is complicated but interesting set of activities. It is like laying out a chess game. A combination of different pieces, moves, and strategy can generate some astonishing results when using the powerful tools like direct sales force, channels, and solution partners. It takes more art than science where to use the knight, when to move the queen, and how to protect the king. Once the ecosystem is set up, the product manager can sit and watch its product's revenue multiply. It is the reward time.

Remember:

- Whereas the overall marketing strategy determines *what* products and services a company will offer and *who* it will offer those to, the go-to-market strategy deals with *how* that will actually happen.
- Just like one size does not fit all, one sales model does not suit all companies. There is a choice of choosing among the direct sales, channel sales, or a hybrid direct touch model.

- The value-added resellers are generally good way to bring the business up in remote smaller geographies especially where local relationships are valued.
- The distributors can unburden a vendor from maintaining its extensive warehouses and distribution network.
- A channel conflict is a situation in which more than one channel partners try to compete over same sales opportunity, trying to sell the same set of products to the same customer.
- Driving business through channels successfully requires a full channel support and reward infrastructure in place.
- Strategic alliances are formed when two companies have their sight on more long-term and much larger objectives.
- In terms of control and flexibility in pricing, positioning, and go-to-market, the OEM may be the best alternative to otherwise building or acquiring a product.
- The solution partnership is usually a result of build-versus-buy-versus-partner analysis to bridge some sort of gap in a company's offerings, but it fits the criteria that the total is bigger than the sum of its components.

CHAPTER-9

PRODUCT MARKETING AND EVANGELISM

Marketing the Products

As discussed earlier, building a product is not enough by itself. Target customers must find out about what the product is and what it is capable of in order to buy it. Therefore, the first step is building the *awareness* about the product. This is marketing 101. If there is no awareness, there is no marketing, there are no sales, and hence there is no business. This chapter is focused on different aspects of the *product marketing* as compared to the overall marketing, since the product marketing is closely related to the product management, and sometimes it is also under the charter of a product manager. In decent-sized companies and especially companies who take product marketing seriously, it is a separate, focused role under the domain of a *product marketing manager* (PMM).

The marketing manager is responsible for promoting and evangelizing a product or product family and providing it the due attention. In this capacity, the marketing manager works closely with the product manager and the sales team. The marketing manager would generally not engage in the concept or execution phases of the product life cycle until before the launch phase, since one of its primary responsibilities is to coordinate or manage the product launch. The product marketing activities continue post launch phase throughout the sustaining life of the product until it is terminated. Regardless of who performs the product marketing activities, let us first familiarize with some of the important functional groups that need to be commonly engaged for the product marketing-related activities. Please note that not every company has all of those functional groups necessarily and it could vary.

Product Marketing

The product marketing team is responsible for marketing the products versus marketing other things such as the company brand. Its primary job is evangelizing and promoting one product at a time. This includes activities like managing the product launches, creating the product messaging, originating the product collateral, as well as planning and participating in public speaking and events to evangelize and promote the products. The product marketing also deals with the industry analysts and media briefings. Another activity that the product marketing usually participates in alongside the product management is the executive briefings, customer tours, sales and partner conferences, and other such activities. The product marketing cannot function effectively unless it has strong support from some of the key functional groups within and outside of the marketing organization. Those functional groups have defined charters and support the product marketing directly or indirectly. In some companies, those may not be separate functional groups but rather just separate functions, some of which could be handled by the same person or group. Some of those functions are discussed next. The rest of the chapter will continue the discussion focused on the product marketing and its key responsibilities.

> *The product marketing is focused on marketing the products specifically.*

Corporate Marketing

The corporate marketing is mainly responsible for the marketing activities at the overall company level, such as brand awareness and elevation, analyst and media relations, press releases, company logo, corporate level messaging, event and trade show management, and other such tasks. There are generally two other entities that are integral part of the corporate marketing organization. One of them is the *analyst relations and public relations* (AR/PR) team responsible for relations and impression building outside of the company. This includes tasks such as conducting analysts and press briefings, issuing press releases, giving awards and recognition, putting together customer case studies and references, tracking the market share and analyst rankings, etc.

The second team is the *marketing communications* (marcom) that is responsible for monitoring and correcting any communication destined for outside of the company. This includes corporate level messaging, product marketing content, any literature, or presentations that are targeted for the customers, partners, analysts, or media. The marcom ensures the consistency of the communication and messaging across the board and its technical and political correctness. The marcom is also responsible for all the brand-related activities. It builds and maintains corporate level logos, graphics, and templates to be used for public-facing collateral for brand protection and consistency. It originates the messaging and organizes the events for the brand-elevation-related activities.

Market Intelligence

The *market intelligence* (MI) or also referred to as *competitive marketing* may or may not exist in all companies, but it is a valuable asset to have. The competitive marketing team gathers the information about the competition, their products, strategies, and tactics. The gathered information is used to plan counterstrategies and tactics to win business against the competition and can prove extremely useful for the sales and marketing teams. Other teams such as the product management and product marketing also consume the intelligence gathered by the competitive marketing and use it for improving the products or for refining the content and messaging. The information is also consumed by the executive team as an input into the strategic decisions. The competitive intelligence team also generates what is referred to as the *fear, uncertainty, doubt* (FUD) against the competition that is extensively used by the sales and marketing teams as a tool to influence customer purchase decisions. Moreover, the competitive marketing team is also responsible for monitoring any new product or strategic announcements from the competition and preparing appropriate response to keep the sales confidence high and to guide them on how to respond to the customer questions and concerns.

Technical Marketing

The *technical marketing engineering* (TME) as it is commonly referred is the technical wing of the marketing organization. The technical marketing team is usually responsible for activities such as conducting product demonstrations, competitive product evaluations, product testing and benchmarking, coordinating roadshows, publishing technical reports and white papers, and many other support activities for the product management and sales teams. The technical marketing team plans and conducts the *roadshows* to demonstrate products and solutions to customers. The technical marketing team sometimes has a dedicated set of resources for the *customer proof of concept* (CPoC) that, working with the *sales engineering* (SE) team, organizes the product and solution testing per customer on-demand basis, before a customer finalizes the purchase decision.

Field Marketing

The *field marketing* team consists of worldwide distributed marketing resources to support the localized marketing activities and sales teams across different geographies. Those activities include telemarketing, lead generation, organizing and participation in local events, public speaking, launching geo-based customer promotions, and arranging customer tours. The field marketing team works closely with the local sales team in assisting them as well as with the corporate headquarters, such as the product manager. The field marketing is responsible for the success, awareness, and promotion of the product in a region. The field marketing is also responsible for generating and managing the leads in the region. The gathered information goes into a *sales leads* data base that is later on triaged by the marketing team to isolate real potential opportunities from not important ones and handed over to the sales and business development teams to be followed up with. The leads are usually marketing, telemarketing, or sales qualified. A useful lead includes the customer contact information, the need, and that there is a budget. The collected leads are *scored* and *nurtured* as they progress. The *lead generation* is a vast topic in itself and is beyond the scope of this book.

Solution Marketing

The *solution marketing* is responsible to define the overall solutions and marketing them along with the product marketing team. The role of solution marketing is very important. As discussed under the go-to-market chapter, most of the products will become parts of some sort of overall solution or ecosystem play with the solution or alliance partners, or it must be so in order to land serious deals and provide customers end to end solution for their requirements. First of all, the solution marketing team defines the solutions along with partner products and offerings to address a particular target market vertical, for example, a mobile clinical solution for the health care vertical, or a supercomputing solution for the oil and gas vertical. The solution achieves certain well-defined goals and has well-defined scale, performance, or other parameters.

Next, the solution marketing team produces the *reference architectures* (RA) that provide detailed architecture and integration details on different components of a solution and how they are qualified, as well as any recommended product configurations that can be deployed. The solution marketing team also produces the *solution briefs* (SB) as introductory marketing material for evangelizing the solutions. Before the solution marketing could publish reference architecture or a solution brief, of course it has to also test and qualify the solution in-house or with the solution partner's participation to make sure that it will work as to be advertised. Therefore, the solution marketing team usually maintains or works with a *solution test team*, which could be a function of the solution marketing team itself or that of the technical marketing or system engineering teams.

Service Marketing

As the name suggests, the *service marketing* team focuses on marketing the services as compared to the products. We will explore services topic in more detail later in the book. A service is an intangible entity. It cannot be seen or touched but can only be experienced. Whereas a product can be owned after someone pays for it, a service is not owned, but rather entitled to. A product can be retained forever, but a service is only consumed at a given point in time. A service is offered either independently just by itself or can be attached to a product or a set of products. There are different types and levels of services in the hi-technology industry that can be offered to the customers. Those services can be divided into basic and value-added services. Simply put, the basic services are almost always provided with the associated products sales. The value-added services are optional and usually take higher degree of skill level to staff and provide those services. They also not need to be associated with product sales and customers can just pay for the service they need.

Just like the products, the services need to be marketed as well, or no one will find out that company actually offers them and what value they offer; hence, no revenue will be generated out of them. Generally, the products and service product management and marketing teams coordinate together on activities such as service creation, pricing, launch, and marketing. The service marketing team can create content focused on marketing services such as services data sheets and other marketing collateral. The team participates in customer executive briefings and other activities to explain to the customers and partners its services offerings and the value proposition. We will explore more on the subject of services in the following chapter.

Legal Counsel

Every technology company has a handful of corporate attorneys that comprise the legal counsel of the company. The legal team guides the company through its day-to-day sensitive matters such as intellectual property, code of business conduct, hiring and firing aspects, business contracts, employment and labor laws, lawsuits, and more. In the marketing context, generally the legal team's involvement is fairly limited to reviewing any important pieces of communication or content that is destined for outside of the company and providing any legal counsel on cautions, disclaimers, and preventive measures. Issues such as publishing legal disclaimers and confidentiality notices in public-facing content, trademark and copyrights notices, legal verbiage, legal contracts such as the resell and OEM contracts, and any terms and conditions in customer promotions, etc., are few common examples of the legal's involvement in the marketing activities.

It is a fine line though that how much paranoid a company's legal team needs to be before it starts becoming a hurdle in doing business. Generally, the role of the legal team is advisory and not enforcing as far as the product marketing activities are concerned. That is, advising the marketing team and product managers in the above-mentioned matters and not directing them to exactly follow what legal team thinks must be the way. Sometimes, the legal team starts micromanaging and policing the marketing activities, making it too difficult for the marketing and sales teams to do what they are chartered to do. Mature companies manage this balance well; therefore, the involvement of the legal team into everyday matters is limited. The product manager and marketing approaches the legal counsel only if they think they need to.

Marketing Program Management

In most companies that take the product marketing and evangelism seriously, launches and other product marketing activities are centrally managed by a *marketing program or project manager* who plans and coordinates all marketing activities and brings different functional groups together as discussed above. This brings order to the overall launch process. The program manager and the marketing manager would generally not engage in the concept or early execution phases of the product life cycle, until during the execution phase and before the launch, in order to start planning for it. The program manager is constantly working coordinating the marketing campaigns and other marketing activities throughout the sustaining phase as well. The program manager also manages and tracks the budget and spending on marketing activities under the *marketing development fund* (MDF) along with the rest of the marketing team.

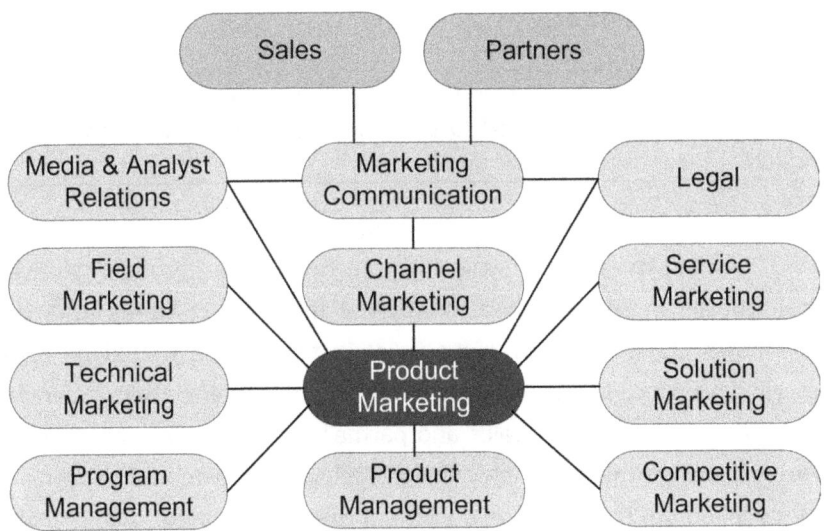

Diagram 1: A Product Marketing Manager's Interaction Scope

Product Marketing Responsibilities

As discussed above, there are several product marketing activities and deliverables that are used to generate product awareness. Most of them were also covered earlier under product launch phase and discussed here in more detail since this chapter focused on the product marketing.

Product Messaging

One item that serves as the foundation for almost all marketing material and public speaking activities is the product-related *messaging*. The product messaging conveys the key message about a product and its main strengths that the product manager wants people to remember and associate with the product. The messaging combines the most important points about the product into a concentrated and concise form. Developing good messaging is more of an art than science and possesses the following qualities:

- It is specific and provides concrete and quantifiable facts about the product.
- It links the technological innovation and strengths with business advantages.
- It is simple and easy to understand.

> *The product messaging conveys the key message about a product and its key strengths that are intended to be remembered and associated with the product.*

Having product messaging nailed down before the product launch is critical, but it can be optimized throughout the product life as the market requirements change. At any given point in time, there should be a clear messaging guidance provided by the marketing manager so that all marketing material can be based on it, sales and partners all speak the same language, and everyone tells the same story. For this reason, there should be formally a *messaging and positioning guide* document drafted by the product marketing that can be referenced for developing the marketing collateral as well as to train the sales and partners. Messaging provides a common denominator, which results in consistency in articulating the product differentiation and value proposition no matter which piece of marketing collateral a potential customer or analyst refers to.

While the product manager may not be the originator of marketing material, but it must be the key party in providing input into defining the *messaging framework*. The messaging developed without the product manager's input is commonly found to be ungrounded, weaker, and factually off, since no one else understands the product better. The quantifiable claims made in the product messaging should be backed up through benchmark testing or other verifiable data points such as competitive data sheets. Example of good messaging could be:

> *The product XYZ provides 6X the performance of comparable products while it cuts down the power consumption by half, resulting into maximum return on investment and lower total cost of ownership.*

In contrast, bad messaging is generic, vague, and provides no real quantifiable facts. For example:

> *The product XYZ provides highest performance and lowest operational costs compared to other similar products.*

While it may sound good, it is perceived as "me too" statement, and many educated customers pass right by it without getting impressed or curious. Smart customers like facts and figures that are measurable and verifiable.

Without the messaging, there is not much to be communicated. The messaging is required to highlight clear differentiation against the competition, which means highlighting why and how new product is better and how much better. Good messaging starts with a good *messaging framework*, which links apparently heterogeneous attributes together so that connections among them could be established in a sensible way. A good messaging framework generates messaging for different levels that resonates with different types of target audience or readers in a customer organization.

The author has pioneered a method of *layered messaging framework* (LMF) for this purpose as explained below in detail.

The layered messaging framework process starts with picking the product level strengths and characteristics as the bottom-most layer—things, if communicated to the audience as is, may not provide much of the value proposition. For example, "10 or 40 GbE capability" or "less than 1 microsecond latency" is not a message but product characteristics that may also be possessed by competitive products. Next, a layer above, the key *technology enablers* are listed that rely upon the product characteristics underneath. This is where the technical differentiation starts taking shape. For example, "6X the performance, 10X the density, 5X the consolidation. 1/4 the latency." This layer is designed to resonate with the technical decision makers on the customer side, such as technical architects and network administrators. It still does not get a CIO's or a CEO's attention. Next, the technology enablers are linked with what *business advantaged* they help generate. This is the third layer, which basically provides a common sense message that can be easily understood by the executive level audience such as a CIO. For example, "150 percent reduction in total cost of ownership (TCO), 200 percent better return on investment (ROI), and 99.999 percent service up time" are clear business advantages that the C-level executives understand.

Finally and optionally, the top layer customizes the whole messaging to industry specific applications that can resonate with the topmost decision makers such as a CEO, the board members, or the investors. It links the business advantages to the end user applications and how the product helps accelerate those unique applications. For example, if it is high performance computing (HPC) network serving oil and gas industry, the application level messaging such as "cutting resource exploration time by half" will click right away with the key executives and shareholders and will get their attention because this directly translates into profitability for them.

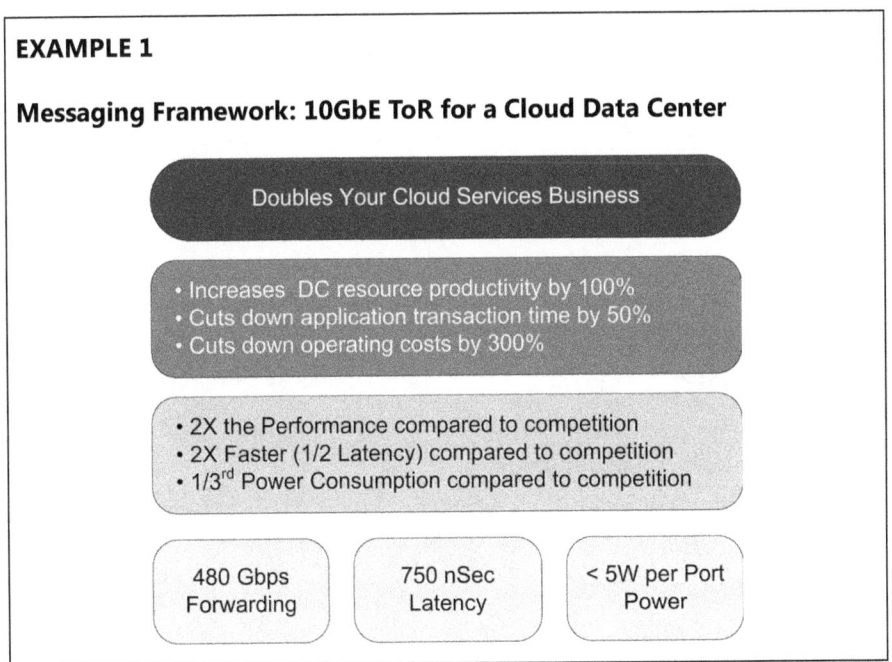

Diagram 1: Layered Messaging Framework

EXAMPLE 1

Messaging Framework: 10GbE ToR for a Cloud Data Center

Doubles Your Cloud Services Business

- Increases DC resource productivity by 100%
- Cuts down application transaction time by 50%
- Cuts down operating costs by 300%

- 2X the Performance compared to competition
- 2X Faster (1/2 Latency) compared to competition
- 1/3rd Power Consumption compared to competition

480 Gbps Forwarding

750 nSec Latency

< 5W per Port Power

Product Collateral

Out of all things important, the product collateral has its own significance and is most sought out pieces of information by the customers throughout the product's life. Whenever a potential customer needs information about a company's products or services, they look for the product collateral on the company website or request the hard copies from a sales representative or partner. There are, in fact, several types of product_collateral focused on variety of objectives and target readers. Just like the messaging, the collateral has to be focused on who the reader is.

It is the product marketing team and the marketing manager who continuously produce useful and high-quality content that is consumed by the sales teams and the channel partners as well as the end customers at all levels. In the hi-technology industry, it takes creative, technical, and analytical thinking to express very complex concepts in a simple to understandable manner on paper. A great marketing manager needs to have a good balance of technical and artistic skills. This rather rare combination of left and right brain thinking is what makes an outstanding marketing manager, producing outstanding and striking content. While there are usually other resources at the disposal of the marketing manager who can write the content, the marketing manager basically originates the ideas, outline, and framework of the important content as well as reviews and approves it before it gets published.

For the hi-technology products and in particular the network products, some of the common product collateral that is developed by the product marketing team may include:

- Product landing web page
- Product data sheet
- Product brief
- Solution brief
- Technical brief
- Customer presentation
- Technical presentation
- Benchmark report
- Product video
- Social media
- Customer case study

Product Web Page

With exploding popularity of the Internet for finding and exploring information, the product web page serves the most important purpose in providing firsthand information about a product. Many technology companies have multiple paths to find the product information through the product landing page. Commonly, every company has a "products" link on their website that provides navigation by product families as a tree structure and a "solutions" link that provides navigation by target solutions and how different products could fit into those use cases. The product web page serves as the central place for holding important information for customers and partners, such as product overview, specification, data sheet, solution briefs, white papers, and other documentation. The product web page or the landing page needs regular maintenance in terms of updates and messages as the product is evolved. There should be consistency between the landing page content and the content in the product data sheet at all times. Usually, the product marketing team tracks the number of hits on the landing page to gauge the interest in the product.

Product Data Sheet

The data sheet is viewed as the most authentic and credible source of product information by the customers and partners. As the name suggests, the data sheet collects the most important data about the product in a single document. The data sheet generally consists of the product overview, key highlights, value proposition, detailed product specifications, industry standards compliance, and ordering information. Sometimes target applications or use cases are also illustrated in the data sheet. The product data sheet is targeted at the technical decision makers on the customer side, and they usually require it to study the facts and figures about the product and to make the educated decisions. It is sometimes hard to be excused of mistakes made in the data sheet, as customers view it intentionally misleading or fact hiding. It is therefore important that the facts listed in the data sheet are accurate and can be backed up with evidence.

For those reasons, the initial draft of the product data sheet maybe originated by the product marketing, but it is closely reviewed and scrutinized by the product manager and even the core team members. It is also common that the product overview section be drafted by the marketing team while the product specifications section is completed by the product manager. The product manager relies on the help of the core team, such as operations, engineering, and compliance for gathering and reviewing all the facts and figures needed under the product specifications sections. The product data sheet is mostly the only piece of product collateral that is reviewed and endorsed by the internal core team. The length of the data sheet depends on the type of the product and how much data there is to list. A good data sheet is not too long to be overwhelming and not too short to miss important facts. Still, the customers almost never read the complete data sheet but rather refer to the sections they are interested in.

EXAMPLE 2

Product Data Sheet

Z-Series 10GbE Top-of-Rack Switches
Enabling Next-Generation Cloud Data Centers

Highlights

- Wire speed 10GbE operation
- Low 450 nSec latency
- 480Gbps forwarding performance
- 480Gbps stacking bandwidth with up to 8 units
- 40Gbps uplinks for consolidation
- Software Defined Networking
- Advance Data Center virtualization
- Storage convergence
- Rich L2-3 feature set
- High scale with 0.5M L2-3 entries
- Only 3.5W per 10G port
- Data Center class high-availability

Data center storage and traffic is growing exponentially, and the bandwidth-greedy multimedia applications are consuming more bandwidth. East-west traffic among data center resources is growing due to more interactive applications. Surge in mobility and smart devices such as cell phones and tablets are also raising the north-south traffic and bandwidth utilization in and out of the data center especially in the cloud environment. Server virtualization and I/O consolidation is multiplying the bandwidth use as most servers, and storage arrays are migrating to 10 GbE already putting backpressure for higher bandwidth uplinks such as 40GbE to the core of the data center.

Above challenges demand higher speeds and feeds in and out of the data centers as well as higher bandwidth network connectivity to the resources (servers and storage arrays). The reduction in IT spending and increase in the cloud based offerings put downward price pressure on the price per port. Furthermore, the rack space is getting limited to be allocated to the networking

---

gear, and there is need for higher densities in least footprint. The Cost/Mb pressure and the profitability need continues to demand higher power and cooling efficiencies for lower operating costs and hence lower total cost of ownership (TCO). The need for more efficient network is here. Ethernet is becoming the choice for connectivity within and inter-data centers due to it economies of scale and simplicity.

The new Z-series high-performance Top-of-Rack switch meets all of the above needs and enables the cloud operators with higher return on investment and better customer experience. The Z-series is designed with wire speed 10GbE for high-bandwidth server and storage connectivity. Its horizontal stacking capability of up to 8 units with 480Gbps bandwidth enables most optimal switching for the east-west server to server and server to storage traffic while 40GbE uplinks ensure high-speed path for north-south traffic. The low latency switching of only 450 nanosecond provides the fastest response to the waiting applications and makes the data center resources look like if they were directly connected. The Z-series packs all this into a compact 1 RU size and only 3.5 Watts per port power consumption that makes it least expensive to operate. In fact, the Z-series could increases data center resource productivity by 100 percent, cut down application transaction time by 50 percent, and cut down the operating costs by 300 percent.

On top, the Z-series helps build a fully programmable network using *OpenFlow* and any ONF standards compliant SDN controller. This plug-n-play capability combined with advanced virtualization features enables the cloud operators to orchestrate and provision advanced services from a single point of management.

Target Applications

Data Center Top-of-Rack
The Z-series primary application is aggregating the server and storage within data center racks using dual-speed 1 and 10GbE links. The Z-series offers both fiber and copper cable connectivity options depending upon the NIC or LoM equipped servers. In this application, a pair of ToR switches dual-home every rack server or storage node for complete service and application availability. The ToR switches then connect to the core in dual-homed way through 40GbE links using short- or long-range optics.

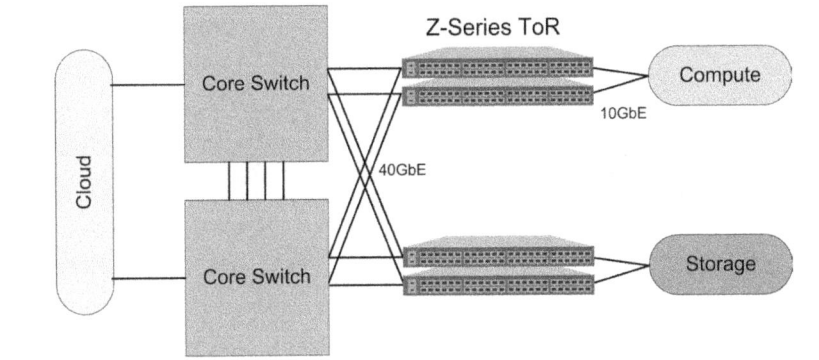

Technical Specifications

Performance	General Specifications	Physical Specifications
• 1280 Gbps switching capacity • 950 MPPS forwarding rate • 450 nanosecond latency • Cut-Through switching • 480Gbps stacking bandwidth	• 48 10GBaseX SFP+ ports • 48 10GBaseT RJ45 ports • 4 40GBaseX QSFP+ ports • 1 10/100/1000BaseT mgmt. port • 1 RJ45 RS-232c console port • 64-bit MIPS processor • 1GB SDRAM • 1GB Compact Flash	• Height: 1.75 inches (4.4 cm) • Width: 18.0 inches (45.7 cm) • Depth: 20.0 inches (50.8 cm) • Weight: 16.0 lb (7.3 Kg)
Feature Scale	**Operating Specifications**	**Environmental Standards**
• 4,094 VLANs • MAC Addresses: 512K • IPv4 Host Addresses: 96K • IPv4 LPM Entries: 512K • IPv6 LPM Entries: 256K • Multicast Entries: 256K • 9216 Byte maximum packet size • 8 QoS queues per port	• Operating temperature: 32-113 F • Operating humidity: 10-95% • Operating altitude: 9,850 ft • Operational shock: 30m/S2 • Random Vibration: 3-5000 MHz • Acoustic Noise: 55-68 dB (A)	• EN/ETSI 300 019-2-1 v2.1.2 (2000-09) – Class 1.2 Storage • EN/ETSI 300 019-2-2 v2.1.2 (1999-09) – Class 2.3 Transportation • EN/ETSI 300 019-2-3 v2.1.2 (2003-04) – Class 3.1e Operational • EN/ETSI 300 753 (1997-10) – Acoustic Noise • ASTM D3580 Random Vibration Unpackaged 1.5G
Media Access Standards	**EMI/EMC Standards**	**Warranty**
• IEEE 802.3ab 1000BASE-T • IEEE 802.3z 1000BASE-X • IEEE 802.3ae 10GBASE-X • IEEE 802.3ba 40GBASE-X	• North America EMC for ITE • FCC CFR 47 part 15 Class • ICES-003 Class A	• 1 year for Hardware • 90 days for Software

Ordering Information		

Part Number	Product	Description
101	TOR-10G-48F	48-port 10GbE SFP+ ToR Switch
102	TOR-10G-48C	48-port 10GbE RJ45 ToR Switch
103	TOR-UL-40G-4	4-port 40GbE Uplink Module
104	TOR-STACK-480	480Gbps Stacking Module
105	TOR-L3-LIC	Layer 3 Feature License

Product Brief

A *product brief* is an overall product guide that provides a brief overview of overall product portfolio and offerings of a company in a short booklet form. Sometimes a *quick reference guide* (QRG) is used to provide an even brief, side-by-side overview of the products and their specifications on a single sheet of paper. Whereas a product data sheet is usually useful for the customers who are looking for more deep down information about a product, a product brief or a quick reference is usually useful as the firsthand overview for customers not yet fully familiar with the products and offerings of a company. Target audience of the product brief is an executive as well as the technical staff on the customer side.

Solution Brief

A *solution brief* is usually a one or two pages document focused on a particular industry vertical solution or application. It explains the key requirements of the solution and how the product and other components of the solution fit those requirements. As we discussed the importance of the solution partnerships and ecosystems in the previous chapters, it should be clear why the solutions and solution-based content are important for a company's go-to-market strategy. The solution briefs are popular with customers as they are short, and they provide the insight into much more relative and real applications for them. Target audiences of the solution briefs are the executives and technical staff on the customer side.

Technical Brief

A *technical brief* is a short document that is focused on a particular technology or feature and how it helps solve certain challenges. It can also be focused on a set of technologies that work and get deployed together as a meaningful technology solution. Usually, the technical briefs are used to market certain proprietary technologies and innovations that the company has developed or to influence the customers to use certain technologies it wants to promote in response to the competition. There could be many reasons for doing so. The main idea is to bring different pieces of technologies together to provide a real-world application so that it builds the understanding about what is the value proposition and usefulness of those technologies, not appreciated otherwise. The target audience of the technical brief is the technical staff on the customer side.

Customer Presentation

A *customer-facing presentation* consists of set of PowerPoint slides that is used to pitch the product to customer executives such as C-level executives. It is a product overview with key messaging and differentiation, which the sales team and partners can leverage to present to their potential customers. The customer presentations are used routinely in the *executive briefing center* (EBC) for customer meetings. The executive briefings involve inviting the customers on-site and showcasing company's products and solutions to them and providing briefings to them. The customer presentations are popular piece of content and extensively used by the sales teams and partners for customer meetings.

Technical Presentation

A *technical presentation* consists of set of PowerPoint slides that explain the product architecture, technical features, and functionalities in deeper details. The technical presentation is focused to cater more technical audience on the customer side such as the network architects and system administrators. It is also used internally for training the technical resources such as the system engineers (SE) and technical support (TAC) engineers.

Benchmark Test Report

The benchmark test reports, as discussed earlier in the book, help build the product's credibility and can be good marketing collateral. To build the new product's credibility, it is sometimes useful to conduct the product testing through an independent party. The third party can put the product through demanding testing and endorse its differentiating performance or functionality. This could be regarded credible by the potential customers because it is based on a "neutral" party's statement. In addition, most of the time, the independent agencies test the product against multiple competing products and given that a product does well comparatively, provides strong evidence to support its value proposition. This helps in building the customer confidence in the product and can be an effective sales tactic. The customers then perceive it as a serious product rather than just another product. The benchmark testing can be expensive, but its impact can last for quite some time. The target audience is usually the technical staff on the customer side.

Product Videos

With the rise of YouTube and other video content-based websites, interest in video has exponentially grown. This has been further accelerated by the new generation of smartphones and tablets where more people prefer to watch videos than reading a paper. The product marketing team can particularly benefit from these trends to market the products. More and more product vendors are choosing to put a *video data sheet* along with the traditional paper data sheet on their web pages. A video data sheet is a few-minutes-long video in which a product manager presents and talks about the key features and capabilities of its product.

Anyone new to the product can watch the short video about the product first and then click on other content to read further if interested. This attracts more customers to learn about the product compared to traditional paper data sheets. However, the video data sheet is not a replacement of the paper data sheet and should not be considered so. The video cannot cover every detail about the product such as specifications, dimensions, standards compliance, orderable part numbers, and so forth. It complements the paper data sheet and other content rather than replacing it, and this is the objective that the product marketing should try to meet when developing the product video.

In addition to the video data sheet, there can be other types of videos developed and uploaded for marketing the product. For example, videos of product showcasing at the industry events, videos from the customer tours, interview-style interactive videos with company executives and key industry analysts, and the video with customer testimonials and experience about using the product could be some of the great tools to market. Existing customer testimonial videos can be very powerful tool to influence other potential customers. The customers tend to trust each other a lot compared to a marketing pitch from the vendor company. Having loyal customers say few good things about a product builds credibility for the product and the company and helps win new business. The target audience is all levels on the customer side.

Flash Animations

In addition to the videos, interest in animated, video-style illustrations is also on the rise. Animations use tools like Macromedia Flash to create self-playing presentations that can express complex concepts as sequence of choreographed illustration and images narrated with a commentary. Those animations are usually short and sweet. The flash animations can be a great marketing tool in the hi-technology world since there is usually product and solutions-related complex concepts that need to be presented in a simplest possible way to the potential customers. For example, there could be a flash animation explaining how a certain industry vertical is evolving, what types of solutions are needed, and how the product being offered fits in there to help solve those challenges. While these days, capturing a video, editing it, and uploading it on a social media website is pretty easy process; comparatively, developing professional grade flash animations takes time. The process of building a storyboard, scripting the animation, narration, selecting images, building animations, and recording the voice-overs takes time and effort. The target audience is all levels on the customer side.

Social Media

With the explosion of social media such as Facebook, Twitter, LinkedIn, and other, more and more people are spending time online to seek and share information. This opens up a huge opportunity to market the products and solutions online and reach masses of people simultaneously. The product marketing team can put a page on Facebook to share the news, industry events, customer testimonials, products and solutions videos, and other interesting content to spread the word. Customers and partners can comment on it, like it, and share it, starting an interactive sequence of feedback and response rather than only one-way marketing. The product marketing team could tweet important news and highlights as well.

In addition, putting up and participating on well-known blogs online is a good strategy. Articles related to the topics of interest to the industry and customers are published on the blogs and can generate interesting debates, which attracts the readers to keep coming back. Advertising on the industry-specific, popular web pages and content web sites, where most of the customers come to seek information, is another marketing strategy. It helps build brand of a company and its products to those who may not know about it. The Internet and social media present huge opportunity and vast marketing channels that can be exploited by the marketing teams to reach potentially millions of eyes globally, without maintaining regional marketing teams. Sky is the limit on how creative and successful a product marketing team could be in benefitting from it.

Customer Case Studies

The customer *case studies* can be a powerful marketing tool to build the credibility fast. A case study is a piece of content written about one customer at a time as a success story. It explains who the customer is, what their business needs or technology challenges were, why they selected the company and its products, how those products or solutions helped solve their problems and improve their business, and what has been their overall experience. In this way, a case study captures a real customer win from start to finish. This can be extremely useful for other potential customers in the similar line of business who are perhaps facing similar issues and shopping for solutions. By reading the case study, they can feel comfortable to trust the product that if it could solve the problems of another similar business, it could perhaps solve their problems too. This motivates them to explore more about the products and solutions being offered and get in touch with the company's sales representative.

Media and Analyst Briefings

One of the most important activities the product marketing assists with is the analyst and media briefings. As product announcement is on the way out, the analysts and media relations team starts lining up the media and industry analyst briefings regarding the new product introduction. Those briefings are opportunities for a product or marketing manager to tell the product or solution "story" in a way it should be conveyed.

The media companies generally include industry-renowned online and paper magazines, blogs, and industry expert editorials. In the networking industry, for example, Network World, Information Week, TMC News, Internet News, and many others. Media companies write articles with their own views and reviews the new product, latest industry trends, and how effective the product could be. The media companies generally set up an interview-style conversation in which the product or the marketing manager is usually the leading personality providing the product overview and differentiation. The editors take notes and then publish articles with their own views mixed in it. This passes through hundreds of thousands of readers' eyes and generates good product awareness. Sometimes there is a live face-to-face interview that is later on broadcasted as a video.

The industry analyst firms generally include the market research companies who provide deeper insight into latest industry trends and projections. They analyze the new products and technologies under broader competitive landscape and usually provide their own expert opinion and analysis on how effective the new product may or may not be. The analysts are well educated in their field and maintain close contact with vendors and customers so they have pretty well-rounded understanding of the trends, issues, and solutions. Some examples in the networking industry include names such as Gartner, Forrester, IDC, Current Analysis, Dell'Oro, Infonetics, and others. The briefings to the analysts are very similar to the briefings to the media companies.

The primary purpose of the media and analyst briefings is to generate positive perception and influence regarding new product in the industry. It is important to note that before those briefings can take place, some basic information about the product must be thoughtfully put together in simple, clear, and concise form for the target audience, usually in the form of a PowerPoint presentation that has a flow and builds up a story. The main parts of a typical analyst presentation may consist of:

- A problem statement
- Solution and new product introduction
- Messaging, differentiation, and value proposition
- Pricing and availability

The problem statement should summarize what was the original problem or set of problems that the product was designed to solve. This could be taken from the original concept proposal. The solution overview and the product introduction provide a brief overview of how the new product solves those problems and what are its main attributes and differentiating features. This articulates the key product advantages compared to other competitive solutions available in the market. This is conveyed using the product messaging developed earlier. Toward the end, the product target availability time frame and the preliminary pricing information are provided.

A good practice is to always seek feedback on the spot from the media and analysts on what they think about the new product and its value proposition. This feedback can be a valuable source for the product manager and marketing teams to refine the product messaging and tune up the product roadmap. But importantly, if there is anything that could have been misunderstood or misinterpreted, the presenter still has a chance to correct it before it is widely distributed.

Public Speaking

The product marketing is not limited to the above-mentioned activities only. The marketing manager participates in the industry events and professional conferences and represents company's vision and solutions to the participants. Examples of such events includes external events such as speaking at a conference on a selected topic, speaking at a customer- or partner-focused private event, speaking on a customer tour, etc., as well as internal events such as speaking in sales and partner conferences and speaking during the sales-focused webinars and trainings. Additionally, the sales team routinely requests the marketing manager's participation in the customer meetings to present the product pitch. This focused evangelism generates awareness about the company and its offerings and helps elevate the brand. In this role, the marketing manager is *the* face of the company for the outside world. Therefore, the personality of the marketing manager, his knowledge, confidence, body language, ability to deliver the content and being a good public speaker do matter.

Product Showcasing

Usually, the press release announcing the new product is aligned with a major trade show or conference event. Companies usually plan for presence at major trade show events every year by placing a booth or having a public speaking opportunity. This provides an opportunity for a company to display their latest and greatest products and solutions to thousands of visitors including potential customers, partners, media, and analysts. Interested customers stop by at the booth, attracted by free giveaways or other incentives, and ask questions. They are usually provided with the product collateral for study while their contact information such as name, title, type of business, etc., is captured as potential *sales leads* for later follow up.

The gathered information goes into a sales leads data base that is later on triaged by the marketing team to isolate real potential opportunities from not important ones and handed over to the sales and business development teams to be followed up with. The trade show and event participation is not always as effective, especially for smaller-size companies. Showcasing a product at a major trade show can be very expensive and usually affordable by the large companies with sizable marketing budgets. How much useful lead generation can be generated out of the trade shows is a question mark, and some companies chose to spend that budget on more focused and exclusive customer events where they can have full attention of the customers.

Showcasing the product is not limited to trade shows only. Companies also showcase the products in their *executive briefing center* (EBC) area, which is maintained at the company's own location for inviting customer executives and decision makers on-site.

The executive briefing center focuses on providing the corporate, products, and solutions overview to the customer executives who the sales team brings in for the visit. The product manager, product marketing, and the sales teams are usually the key participants in customer briefings. The visiting customers are presented with the product overview followed by a tour of the product showcase area where the company products are displayed. Another way to showcase the product is through the planned customer tours. In this case, a company could come up with creative ways, such as putting together a mobile EBC in a large vehicle that goes to the customers at their locations instead of inviting them over. Executive briefings generally result into rather long-term sales returns.

Yet another way to showcase the newly launched product is to reserve certain number of product units for the sales and the partner use, which can be distributed worldwide to become part of the localized *product demonstration pools (demo pools)*. The sales engineering team maintains those demo pools and rotates the product around for showcasing and demonstrating it to the potential customers through a *request-and-reserve* process. This provides a personalized experience to potential customers. The new product can also be part of a *try and buy program* in which case a customer has the opportunity to try the product in its environment before buying it. The try and buy program works well for very expensive products where customers may be reluctant to trust the new product and are not willing to pay for it without having complete satisfaction about it.

Customer Tour

Another activity that may or may not always happen is a preplanned customer tour. The marketing team selects significant potential customers who are on the *must-win* list of the company and plans regional multicity tours visiting them at their locations to build the product and brand awareness. Customer tour may happen with or without showcasing the products. The product manager may or may not accompany the tour. Although customer tours may not translate into immediate sales pipeline necessarily, it eventually pays off in the long term.

Customer References

Nothing else influences potential customers better than listening to existing customers say something about a product or a solution. Sometimes, larger sales opportunities require that the customer references be provided to the new customers to prove that the product and solution they are planning to deploy is in fact trustable and reliable. This is especially important in the case of new products, new customers, or new companies. The new customer could either demand the details on other customers who have already deployed the product in similar business environments or use cases, or it may ask for the permission to contact and speak to some of those customers to get the feedback about their experience. Not all customers like to be referenced and allow their names or information to be shared publically or privately with other businesses. Customer names, logos, or other details also cannot be used in public-facing presentations or other marketing content without prior permission. Therefore, prior authorization is necessary from the customers being referenced.

The product marketing or the analyst and media relations team maintains a database of the "reference customers" worldwide that it can use when requested by the sales teams. The database includes the details about what level of involvement in the reference process the customer can offer. Using and sharing the customer information is usually not part of a deal under the privacy policies, and doing so could result in legal consequences. However, at the time a product purchase decision is being made by the customer, authorization from the customer to be used a reference or for a case study later on could be requested but should never be stressed upon.

Email and Phone Marketing

Despite the rise of digital marketing and other newer marketing mediums, the old-fashioned email and telemarketing still continues to be effective depending on the type of businesses, industry, products, and demographics. Using periodic email newsletters and updates to existing and potential customers keeps them up-to-date about a company and its products. Even if a fraction of the recipients actually read it, the return is worth it because it is cheaper medium. The art of successful email news flash is that it should be relevant, concise, and attractive. Since these days, email tools allow embedding graphics and videos inside the messages; some really influential email messages can be created and broadcasted to the target customers.

The conventional telemarketing or cold-calling may or may not be as effective these days depending on how it is done. Most of the customers do not like to be cold-called, and it may in fact annoy them. It may not suit products used for serious business applications. However, targeted calling based on the lead generation from sources such as customer events, field marketing, or inside sales could prove useful. In this case, instead of random calling, the right person at the right position is approached. Also, a recommended approach is usually to contact the person by email before calling to provide a heads-up or to set up a time to speak. In addition, the sales representatives are provided with the marketing material and call scripts that enable them to have a very structured and to-the-point conversation. This is more professional approach for having the phone conversations. Having such marketing scripts help the salespeople narrow down the conversation into areas that require further discussions and help them set up a follow-on meeting or next conversation—which is usually the goal of the first conversation.

Sales and Partner Enablement

So far in the book, in the capacity of product launch and product marketing, we have discussed multiple aspects of evangelizing and promoting the products. However, one of the extremely important aspects of making a product successful is how to enable the sales teams and channel partners in making them successful at selling. Only a successful sales team and partner base can help make the products successful. This is exactly this sales enablement role of the product marketing that matters the most. We have discussed that the product marketing team produces lots of content that helps the sales team and the partners in influencing the end customers. This includes product messaging, collateral, and showcasing. The product marketing enables the sales team by putting all the pieces together to pitch the products and solutions to the customers properly and effectively.

Sales and Partner Training

One of the most important activities before launching the product is to train the sales force and the partners on the new product. Generally, this "how to sell" training focuses on the product features, differentiation, positioning, applications, and key messaging as well as any weaker points to tackle. The product manager is the key participant in the training usually or partners with the rest of the marketing and sales resources to structure and deliver the training. The customer- and partner-facing product update and training could be on-site or off-site. It could be organized at leading partner or company locations where partners or customers could be invited to attend. Or alternatively, it could be virtual training in the form of webinars that anyone could conveniently attend from anywhere around the world.

Generally, a good training is segmented in terms of the content focused at two different sets of audience. One part is focused on purely the salespeople such as the account managers (AM), channel account managers (CAM), business development manager (BDM), and other sales representatives. This part of the training focuses on high-level product pitch, positioning, market opportunity and objectives, key messaging, value proposition, pricing, and commercial options. The other part of the training focuses on the sales engineering (SE) and technical resources that in turn support the salespeople.

This part focuses on deep dive product details such as product architecture, features, use cases, configurations, and competitive analysis, etc. Furthermore, the training content should be segmented for the sales team and that for the partners since not every detail can be shared with the partners; and on top of that, partners have some unique training requirements of their own.

The sales training should not be taken lightly; neither is it a one-time task. Longer learning curves are not good thing in sales. Every day a salesperson is spending in learning, it is not being spent in actual selling and hence generating the revenue. Focused, repeated, and high-quality sales training builds competent sales force that can really make the difference. The sales force and partners that are not well trained are mostly confused and come back with lots of question all the time that the product manager or the product marketing has to deal with. This can cause nuisance. On the other hand, a well-trained sales team and partner base can handle most of the challenges independently and make the desired impact.

Sales Playbooks and How-to-Sell Guides

Once of the important pieces of useful sales enablement tool that the product marketing can help with along with other teams is the *sales playbooks*. A playbook captures the best practices and rules of engagement for the salespeople that they can apply against particular competition under specific situations. Those rules and practices are result of hard-learned field and past sales experiences. They are meant to help bring new salespeople and partners onboard but not to replace the conventional sales training. The sales playbooks help create consistency across the sales team and shape the behavior since everyone learns to apply the same rules. It helps the sales teams and partners to determine what the top sales opportunities are for a given product or solution, how to approach and qualify those, how to engage other resources and when, and how to conclude a sale.

> *A sales playbook captures the best practices and rules of engagement for the salespeople that they can apply against particular competition under specific situations.*

In addition, a playbook can be focused on specific market segment or competition. It can help the salespeople educate on typical competitive tactics and traps, their messaging, strengths, weaknesses, and how to handle and respond to those. It guides on how to leverage the product messaging and use the FUD principles against the competition and how to flip a given deal in company's favor. The playbooks, when combined with the other competitive analysis and guides that the product or competitive marketing teams originate, provide extremely powerful tools at the sales team's disposal. The salespeople are fully educated about who the competition is and understand the full *strengths, weaknesses, opportunities, threats* (SWOT) picture about it. Without proper sales playbooks, every salesperson reacts to a situation differently, and the net result is a bit chaos with hit-and-trial sales approach. Sales playbooks bring the organization to the sales process and hard-earned collective knowledge in one place, which helps the sales force break into large and complex opportunities. Discussing the playbooks and the competitive analysis in detail is beyond the scope of this book.

Sales, Customer, and Partner Meetings

Other sales-enablement activities that the product manager and the marketing teams usually help with include participating in the sales and customer meetings to help the sales process, travelling to the customers and partners for product and roadmap presentations, and answering any questions on behalf of the customers on daily basis.

We have already explored the customer executive briefings of such activity. Additionally, there are sales and partner conferences every year, sometimes twice a year, that require active participation of the product management and the product marketing in terms of presentations and training related activities. Moreover, there could be closed forums such as the *customer advisory council* (CAC) and the *partner advisory council* (PAC) that periodically meet to discuss the key issues customer and partners face and to collect the feedback which could prove extremely useful in recalibrating and aligning the company strategy and roadmaps. The direct participation of the product manager and the marketing manager in customer engagements adds credibility and helps the sales process because customers tend to believe that the product managers are honest and knowledgeable. And this perception is the reality most commonly. However, the product managers need to always balance how much time they can spend into the sales-enablement activities without compromising their core responsibilities.

Partner-Focused Marketing

As we discussed different channel sales and partnership models in the previous chapter, among many other channel-enablement activities, one key one is the channel-partner-focused marketing. Driving business through channel sales model successfully requires a full channel support process and infrastructure in place. The channel partners rely heavily on the product vendors for knowledge and skill transfer, training, and marketing support. The marketing development fund can have direct influence on how much a partner is motivated to promote one vendor's products versus another. The partner uses the funded dollars to organize business development and marketing activities such as customer events, meetings, and entertainment activities to generate the business. Even if partners use their relationships and market reach and arrange marketing events and activities, they mostly seek the support from the vendors to provide tailored marketing collateral and public speakers who are subject matter experts.

Usually due to the focus and effort it requires, there is separate marketing team for the partner-enablement activities. We will not discuss this in detail since it's a subject in itself. The main point is that the product messaging and related content developed by the product marketing team for its direct customers that we have discussed earlier may not be taken "as is" and used for the channel partners. Although some of the content could be repurposed, it usually requires customization to make the content useful for the channel partners. In the case of OEM partners, the content may need to be fully rebranded or developed from scratch to suit the partner branding and go-to-market strategy.

Moreover, there are additional set of marketing activities that are required just for the partners. The partner specific communication also needs to be managed by the channel marketing team. This includes regular communication regarding important news and updates in the form of partner focused email newsletters and blasts. The partners focused training and webinars need to be organized and delivered. Usually, the channel marketing team maintains a partner specific online *portal* that every partner has access to. The product content is uploaded to the portal and available for the partners' use. Other important information and announcements such as service notifications, product general availability, pricing and promotions, etc., are also posted at the portal. Therefore the portal needs regular updates and maintenance.

Key Takeaways

It will be really sad to see a great product sitting in the warehouse and not moving since no one knows much about it being so great. Therefore, the product managers love the opposite. They love seeing their products pictured in the media content, people talking about them, and the inventory running out every time they bump up the forecast due to exploding demand. Those kinds of ideal things can only happen as a result of a great marketing team behind the product and a great sales team in front of it. It is lots of fun proudly talking about how the product is better than its competitors in public forums and watching the competition trying to pick on something about it or looking the other way.

Remember:

- Building a product is not enough by itself. Target customers must find out about what the product is and what it is capable of in order to buy it.
- The marketing manager is the face of the company for the outside world.
- The product marketing team is responsible for marketing the products versus marketing other things.
- The product messaging conveys the key message about a product and its key strengths that are intended to be remembered and associated with the product.

- Good messaging starts with a good messaging framework, which links apparently heterogeneous attributes together so that connections among them could be established in a sensible way.
- A good practice is to always seek feedback on the spot from the media and analysts on what they think about the new product and correct for any disconnects on the spot.
- It takes both creative and analytical thinking to express very complex concepts in a simple to understand manner in the product collateral.
- The sales playbook captures the best practices and rules of engagement for the salespeople that they can apply against particular competition under specific situations.
- Video and social media are changing the way marketing is done and opening up global instantaneous reach to target customers.

CHAPTER-10

SERVICE OFFERING

Importance of Services

Most of us know that in today's hi-technology environment, revenue is generated primarily by offering and selling two types of entities: a product or a service. As we discussed in the very first chapter, a product is a tangible item that you can see, touch, feel, and use in a certain way. They are based on "goods." A product is usually designed to solve a particular problem or to serve a particular purpose; it is usually designed for particular needs of a particular set of users who will likely use it. It is therefore designed per the requirements those users have and per the specifications and attributes that those users like the product to have. However, with the growth of Internet, a product now can also be something you cannot touch, but you can still see it and use it in a defined way. Many software applications that are sold and used in the cyberspace are examples of this. A product, therefore, is an entity produced by labor and designed to solve a certain problem, to function in a certain way, and to produce a certain output for monetary purposes and as demanded by the product users.

A service, on the other hand, is an intangible entity. It cannot be seen or touched but can only be experienced. Whereas a product can be owned after someone pays for it, a service is not owned, but rather entitled to. A product can be retained forever, but a service is only consumed at a given point in time. A service is offered either independently just by itself or can be attached to a product or a set of products. Just like the products, in this context services are also offered for generating money. Generally, any work that involves a human interaction on top of what product cannot do by itself is put into the service category and is usually based on leveraging someone's skills or expertise in an area. For example, when a customer has a problem using a product it just purchased and picks up the phone and calls the technical support of the company, it is looking for a service, the very service offering that is tied with the purchase of the product. This service was either already paid for inclusive in the price of the product or is paid for separately. Usually, there are well-defined constraints that are put in place around the service offering so that unreasonable or unlimited expectations are not set.

> *A service is an intangible entity that is offered by someone and consumed by others at a given point in time, usually based on leveraging someone's expertise and usually paid for.*

Although the major portion of this book is focused around designing, building, and marketing the products, the importance that services have these days and the impact they can make on the product success and on the overall revenue and profitability command a detailed discussion. However, this book does not specialize in services management or marketing; therefore, most of the discussion is in the context of services attached to a product or offered on top of it.

Services Product Management

Just like the product-based product management, the *services product management* involves defining, creating, offering, and managing services for monetary purposes. The service product management involves managing service contracts and processes. Since there is no actual tangible product to manage in this case, it is mostly about managing expectations and experiences. It is complicated because there is usually lots of human interaction involved. Just like a conventional product has differentiation based on feature, functionality, and performance, the services are usually differentiation based on the quality of people, their expertise, training, track record, and credibility besides cost advantages. A services product manager identifies the opportunities for profitable services business, designs, and packages the services and delivers them to generate the revenue.

> *The services product management means defining, creating, offering, and managing services for monetary purposes.*

The services product manager performs many activities that are common with the regular product management. This includes for example concept, pricing, positioning, competitive analysis, and marketing activities. The services product manager, however, is more of a participant in the core team activities rather than asking them to deliver something. In this capacity, the services product manager stays up-to-date about the core team activities, in order to plan and create new services for the new products and works closely with the product manager to coordinate the product launch and sustaining as well as the termination phases.

Service Life Cycle

Very much like a product, a service also goes through a full life cycle starting with a service idea definition and concept, service creation, launch, sustaining, and finally the termination phase. Each and every phase in the life of a service is directly supervised by the services product manager.

> The service life cycle consists of a service life as it moves from cradle to grave including service planning, creation, offer, management, and termination phases.

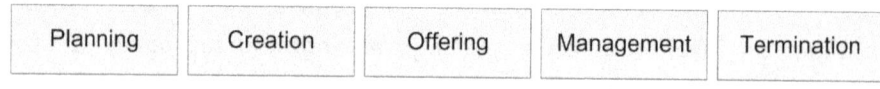

| Planning | Creation | Offering | Management | Termination |

Diagram 1: Service Lifecycle Phases

The service *planning phase* deals with putting together a detailed plan on what, why, how, and when. What does the customer need? What type of service will be created to fulfill the need? And how it should be offered? Those are some of the questions that must be answered in the planning phase as clearly as possible. The planning phase consists of several activities including a service idea, concept, definition, and high-level go-to-market plan. The service offering parameters and scope are defined, and the target set of customers is identified.

The service *creation phase* deals with actual development of the new service. Competitive analysis and total cost of ownership analysis are performed, the new service part numbers (stock keeping units or SKU) are created, the offering pricing is set, and the service contracts are defined. The services, just like the products, are listed on the company's price list and can be ordered. Each service part number on the price list has a list price. Just like the products, the services can be discounted as well and have associated average sale price and margins. Sometimes, a *blended margin* including the product and the attached service is calculated to see the overall profitability picture. We will explore the service-based margins more in the following sections.

The *offering phase* is like a product launch phase. At the service offering stage, a new service is announced or offered to the customers. Just like the product, the new service must be marketed, and target customers should be made aware of it. After a service has been offered, it needs to be managed. This includes making sure that the required services are actually delivered timely when needed by the customers who have paid for them and at the required level of quality. The commitments are made per the *service level agreement* (SLA) in the *service contracts* (SC). It defines *how* those services are to be delivered and managed. Conforming or deviating from the service level agreements will result into good or bad customer experiences.

Eventually, during the *termination phase*, a service will be terminated when its demand fades away and it is no more needed. At which point, it is perhaps replaced with a new service offering. It is also possible that the product a service is attached with reaches the end-of-support stage, and the service has no reason to exist anymore. This has been discussed under the product end-of-life subject in the previous chapters.

Service Attachment

Before a product is launched, all the services that need to be defined to support the product in the field later on must be finalized and created. Those services will be attached to the product throughout its life until the product is terminated, and a milestone called the *end of support* is reached. The way services are attached to the product is either *solid* or *dotted*. The solid relationship means that customers are by default entitled to receive them if they have purchased the product. In other words, those services are already paid for and included in the product price. The dotted relationship means that those services need to be separately purchased and paid for on top, in order to benefit from them.

For example, basic software updates and technical support services may be included with a network product at the purchase price, while the customer may choose to buy additional services such as extended warranty, after-hour or next-business-day service for expedited returns and replacements, or even an on-site service option on top at additional price. In addition, the customer may choose to purchase premium services such as help with installation, configuration, and testing of the product to commission it correctly into its environment. The customer could even buy the ongoing monitoring and optimization services of the customer environment. The *service attach rate* (SAR) tells someone how much of a service is being purchased as a ratio of the associated product. It is one of the indications of the success of the services business. Higher attach rates indicate a healthy services business while the lower attach rates indicate the poor services business.

EXAMPLE 1

Product Attached Service Offering with 10GbE ToR Switch

- Following basic services are to be offered with the new Z-series Top-of-Rack switch
- Only the first service (software and tech support) will be included with the product price

Service Offering	Global TAC 24 x 7 x 365 Days	Access to Online Service Portal	Return & Replacement	Advanced Hardware Replacement	On-site Service
Software and Tech Support	Yes	Yes			
Extended Warranty	Yes	Yes	Yes		
48-Hour After Hours Return*	Yes	Yes		2 Days	
24 Hour After Hour*	Yes	Yes		Next Day	
8-Hour After Hours Return	Yes	Yes		Same Day	
24-Hours On-site*	Yes	Yes		Next Day	Mon–Fri
8-Hour On-site	Yes	Yes		Same Day	Same Day

*refer to the business days

- Service list pricing and orderable part numbers are as follows.
- Similarly the service pricing for other ToR SKUs needs to be created.

Service Part Number	Service SKU	Product Part Number	Product SKU	Service Offering	List Price (US)
1-101	SW-TAC-TOR	101	TOR-10G-48F	Software and Tech Support	$1,050
2-101	EW-TOR	101	TOR-10G-48F	Extended Warranty	$1,250
3-101	48HR-RMA-TOR	101	TOR-10G-48F	48-Hour After Hours Return	$1,450
4-101	24HR-RMA-TOR	101	TOR-10G-48F	24-Hour After Hour	$1,650
5-101	8HR-RMA-TOR	101	TOR-10G-48F	8-Hour After Hours Return	$2,450
6-101	24HR-ON-SITE-TOR	101	TOR-10G-48F	24-Hours On-site	$2,050
7-101	8HR-ON-SITE-TOR	101	TOR-10G-48F	8-Hour On-site	$2,850

Service Margin

Just like the products, services have the profit margins. In fact, services usually have better margins, and the premium or value-added services have even better margins. This is possible because services do not have any cost of goods sold (COGS) for materials or any manufacturing overheads associated with them unlike the regular products that do. The costs associated with service creation are usually minimal in terms of paperwork, etc. The costs associated with service offering are usually the labor-related costs such as employee salaries and training costs, or they could be costs associated with the returns and replacements as well as maintaining the service depots, shipments, and other logistics. The later costs are generally accounted for under the cost of revenue while the former under the operating expenses in the income statement. The service margin directly contributed to the gross margins and net income on the company income statement and helps improve the bottom line.

The service margin provides the net profit picture on per service basis.

The margins for services heavily depend on the type of service and what is involved in creating that service. Generally speaking, the gross margin can be calculated as below:

$$\text{Service Margin} = \frac{(\text{Service Average Sale Price} - \text{Service Creation Cost})}{\text{Service Average Sale Price}} \times 100\%$$

Just like the products, services also have average sale price (ASP) since they can also be discounted and negotiated. And just like the products, the average sale price for services is easy to calculate. The services product manager needs to set a list price of the service that will be advertised to customers on the company's price list. The services product manager also estimates the attach rate and average discount per service. In doing so, the services product manager extensively uses the competitive benchmarking. The discounts could vary based on the geography or market segments. Then the service's average sale price can be calculated as:

Average Sale Price (ASP) = Service List Price X (100% – % Service Discount)

EXAMPLE 2

Service Margin Analysis

- Table below calculates the service margin per service SKU using 60 percent discount.

Service Type	Target List Price	ASP @ 60% Discount	Estimated TAC Cost @ 5%	Estimated RMA Cost @ 2%	Total Service Cost	Margin @ ASP
Software and Tech Support	$1,050	$420	$52	$21	$73	82%
Extended Warranty	$1,250	$500	$62	$25	$87	82%
48-Hour After Hours Return	$1,450	$580	$72	$29	$101	82%
24 Hour After Hour	$1,650	$660	$82	$33	$115	82%
8-Hour After Hours Return	$2,450	$980	$122	$49	$171	82%
24-Hours Onsite	$2,050	$820	$102	$41	$143	82%
8-Hour Onsite	$2,850	$1,140	$142	$57	$199	82%

Service Entitlement

The service entitlement is the right to receive a service as a result of a legal service contract. Per the service contract, the customers are *entitled* to receive certain services, per certain scope, and within certain parameters. The entitlement ensures that the customers get the services they need per service level agreement. For example, the customers could be entitled to receive the twenty-four-hours-a-day, seven-days-a-week, all-year (24x7x365) technical support service, replacement merchandise authorization (RMA) for malfunctioning hardware, access to new software, and access to certain web portals and knowledge base. The customers are usually required to provide certain information to the vendor offering the service in order to authenticate the valid parties entitled to receive the service. This may include providing the product serial number and a reference to the active service contract.

> *The service entitlement is the right to receive a service as a result of a service contract and a service level agreement.*

Type of Technical Services

There are different types and levels of services in the hi-technology industry that can be offered to the customers. Those services can be divided into basic and value-added services. Simply put, the basic services are almost always provided with the associated product sales. The value-added services are optional and usually take higher degree of skill level to staff and offer those services. They also not need to be associated with the product sales and the customers can just pay or the service they need.

The service can also be segmented sometimes into presales and post-sales services. The presales services are the type of services that are consumed in anticipation of a product sale and its deployment in a customer environment to make the environment ready for its adoption. Those services are mostly based of strong planning and readiness. For example, in the networking industry, such services may include site planning, network design, price quoting, equipment selection, procurement, etc. The post-sales services, on the other hand, include the type of services that start right away with acquiring a product, and they continue as long as the product stays in operation. For example, in the networking industry, such services may include technical support, hi-touch support, network monitoring, maintenance and optimization, design changes for future upgrades, etc. Below, we will explore some of the common types of presales and post-sales services.

Technical Support Service

Probably the most common type of service most people are familiar with is the *technical support service* (TSS). The technical support provides the troubleshooting and resolution service to a problem that may be related to the product operation, functionality, or performance. In this sense, the technical support is a "reactive" service. For example, if a customer is having issues starting or operating a product it has just purchased, or if the product is not functioning properly, or if the product performance is not up to the levels advertised, the customer may reach out the vendor company's technical support and ask for the help. During the troubleshooting or problem investigation, the company's technical support expert may find out that there are no issues with the product, but rather the customer could not understand or follow the procedures properly as outlined in the product manual and so forth. In which case, the customer is educated, and the issue is resolved usually right away, unless the product has been damaged and not usable.

On the other hand, the expert may find out that there is indeed an issue with the product, in which case, it follows the company's process for the problem resolution. A standard process at high level usually includes opening up a *case* and issuing the customer a case number for tracking the issue later on, gathering the required data for analysis, forwarding the case to internal engineering teams for investigation, and providing the updates and resolution to the customer later on. The diagram below shows a detailed workflow of how technical issues are usually handled.

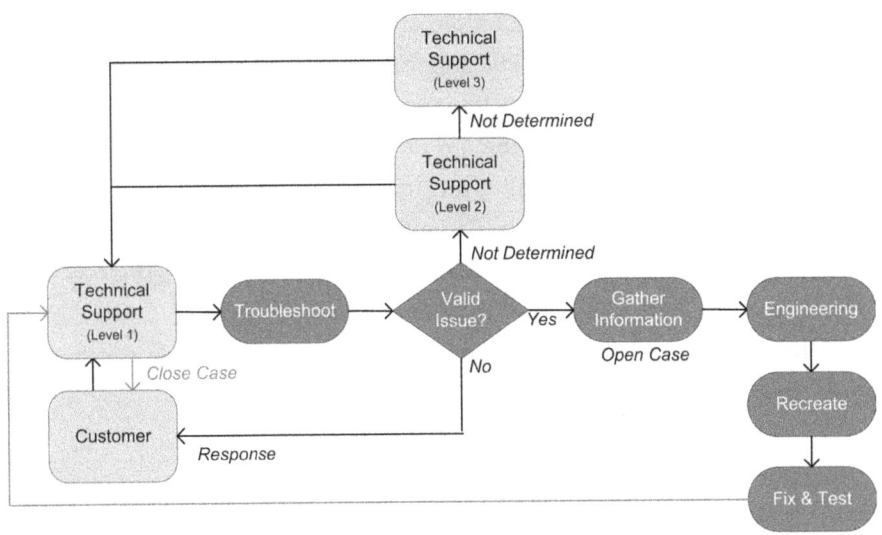

Diagram 2: Typical Technical Support Workflow

The group that generally handles the customer issues and provides the technical support services is also referred to as the *technical assistance center* (TAC) in the hi-technology industry. Within the technical support team, there are usually multiple levels of experts and skill sets, and their time is appropriately allocated. When a customer calls the very first time to report an issue, usually a *level-1* expert speaks with the customers, asks questions, and determines the nature and scope of the issue. The product vendors like to resolve as many issues as possible, as fast as possible, at the level-1 service level, because that is the least expensive. However, there is still a cost associated with handling every customer call. If the expert determines that the issue being reported is a valid problem, it opens a case or *trouble ticket* to track the problem. Every problem has a unique case number for its life. The expert gathers the required data that it needs to analyze the issue later on without contacting the customer and sets the response time expectation with the customer. It then forwards the case along with the gathered data to the internal functional teams such as engineering.

The functional team looks at the incoming issue at high level, prioritizes it in comparison to the other open issues, and assigns an engineer or technical expert to work on the problem. The assigned expert takes a look at the data that the technical support had collected and requests any missing information it may need through the technical support. The technical expert, along with the technical support, may try to "reproduce" the problem in-house—that is, it may try to simulate the environment and conditions under which customer encountered the issue, to establish a cause and effect relationship. Once the root cause of the problem has been identified, only then a fix to the problem can be developed. Based on the investigation done, the expert develops a fix and provides to the customer through the technical support. Most of the time, the fix is in the form of software update or *patch*. However, this may not always be the case.

The customer tries the provided fix and verifies if the issue still persists, or it has disappeared. Sometimes, the original issue disappears, but new issues surface due to the changes made to the product in attempt to fix the older issue. This is referred to as the collateral damage. In which case, the cycle will be repeated. If the issue is successfully fixed and customer verifies it, the technical support expert documents the history, and the case is closed. The problem has been resolved. In case the level-1 expert cannot determine if the reported problem is really an issue or not or the exact nature of the issue, it can escalate the issue to the next level, a *level-2* expert who will troubleshoot the issue and follow the same process discussed earlier or assist the level-1 expert in narrowing down the problem. If the issue is so complex that the level-2 expert also cannot determine the nature of it, it could be escalated to the final level, a *level-3* expert. Due to the level of expertise and skills, chances are that as the problem moves up the levels, it gets nailed down and put on the track for a fix.

The technical support service offering may include different options for the customers to purchase a service based on its business needs. As illustrated in the examples earlier, this may include basic software and technical support service, return and replacement of faulty equipment, and the above services within a time frame that may suit the business urgency, ranging from few business days to a few hours for a response time. This sliding scale service offering helps offer a better customer experience, which results in customer loyalty and long-term customer retention.

Premium or High-Touch Services

In addition to the basics technical or maintenance services, companies may offer on-top or premium services for more personalized attention. For example, such services may provide dedicated resources and single point of contact for the customers. In additional, such services may include "proactive" services on top of basic reactive technical support, such as the best practices guides, recommendations, and optimization of customer environment to increase business efficiency and productivity.

For example, one such service could be the *hi-touch support service* (HTSS) with which some vendors offer on-top personalized technical support for important customer who want exclusive attention, faster response time, and are willing to pay for it. The example of such customers may be the customers who in turn provide services to others using the vendor's products and cannot afford to lose time in case of an issue. For example, in the networking industry, the carriers offering cell phone service could be one such example. Since any issue in the network may cause disruption in the mobile phone service to several subscribers, the carrier cannot afford to wait for the resolution per the vendor's standard operating procedures, and its need assistance urgently and anytime it needs it.

Such businesses may want dedicated technical support personnel on standby for supporting their needs. Those technical experts are sometimes trained and familiar with the customer's environment intimately, so in case of an urgent issue, it takes much less time for them to find out the root cause of the problem. It results in better customer experience since every time a problem occurs, the customer does not need to explain all the fundamentals and waste time in basic investigation and information gathering. Such important or global accounts have their information stored and the support personnel dedicated to them to provide the due attention, service, and resolution they deserve.

A typical premium service offering in a typical networking products based company may look like this:

- Dedicated resources

- o A technical support (TAC) engineer
- o A program manager

- Personalized attention
  - o Periodic customer open technical cases reviews
  - o Reviews for regular maintenance activities
  - o Assessment of issues

- Ongoing proactive maintenance of the customer environment
  - o Periodic customer site visits
  - o Optimization recommendations
  - o Service escalation on behalf of customer business urgency

All of the above options may not be offered to all the customers. The premium services are generally categorized into multiple tiers or options to suit different customer needs and to maximize the service revenues. For example, service offering may be split into three foundation levels such as *silver, gold,* and *platinum,* much like the credit cards or a hotel point scoring system. Depending on the service level, a customer may or may not be entitled to certain type of services.

EXAMPLE 3

Premium Services Offering

- Table below shows the premium services that can be purchased with Nubes Networks products:

Service Type	Silver	Gold	Platinum
Dedicated TAC Engineer	Yes	Yes	Yes
Localized TAC Engineer			Yes
Dedicated Program Manager		Yes	Yes
TAC Case Life Cycle Handling	Yes	Yes	Yes
Customer Environment Assessment			Yes
Knowledge Transfer	Yes	Yes	Yes
Reviews and On-site Support per Year	4	8	12

Professional Services

In addition to the basic or premium support services, some companies provide the *consulting services* (CS), also called the *advanced services* (AS), or the *professional services* (PS). The professional services are usually based on hard to find expertise offered by highly skilled professionals and provide highly tailored services. In the networking industry, for example, those services can range from basic site planning, rack and stack, configuration services to presales network design, equipment selection, testing and commissioning, post-sales optimization and proactive performance monitoring, and network upgrade services. With these types of services, the service experts wear their *consulting* hats on and provide professional advice and recommendations in the best interest of the customer business goals and success.

The professional services are typically focused vertically and deeply into different industry verticals based on a type of business model called *practices*. For example, in the networking industry, the professional services could be based off the practices such as the data center and cloud practice, managed service provider practice, triple-play services practice, and so forth. In this case, the professional services consultant captures the customer requirements and delivers a well-documented solution proposal in response to those requirements. The associated cost estimates are also provided should the customer choose to implement the recommended solution. A typical professional service offering in a networking company is mostly based on the proactive measures and may look like this within a practice:

- Planning services
  - o   Customer site survey and readiness

- Design services
  - o   Customer environment assessment
  - o   Network design and documentation

- Implementation services
  - o   Procurement, installation, implementation, and provisioning
  - o   Testing and qualification

     o   Migration from older equipment

- Optimization services
  - o   Education and knowledge transfer
  - o   Customer environment monitoring and improvements
  - o   Change management and documentation

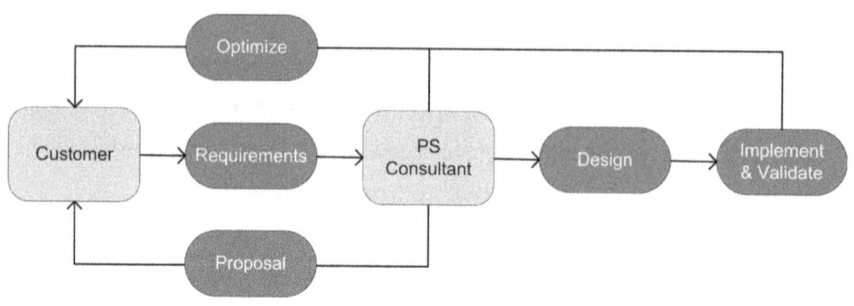

Diagram 3: Typical Professional Services Workflow

The professional services are a high-margin business. The only cost involved usually is the salaries of the consultants involved. The professional services are usually *billed* on per-hour basis just like the regular consulting. For example, in case of the networking products, the *Network Consulting Engineers* (NCE) provide those professional services. Since the consultants are paid a fixed salary, the revenue generated by billing their time on per-hour basis can generate additional revenue that pays for the consultant itself as well as generates gross margins on top. Therefore, the professional services can contribute to the company's top and bottom line significantly. This is one of the reasons many vendors and channel partners like to offer the professional services tied to certain products and solutions. However, creating and offering the professional services is not an easy task. It requires attracting top talent, a brand recognition, deep industry knowledge, proactive services marketing, and sales to make it successful.

Managed Services

Some companies may choose to outsource maintaining their operations to an outside company. For example, in the case of networking industry, a company may find out that it can achieve significant cost savings by outsourcing its IT department instead of hiring full-time IT staff. It may seek, in this case, managed services from an outside company. The managed services may include simply 24x7x365 monitoring services such as the *remote operating services* (ROS) or full end-to-end services including procuring and installing equipment, hosting the business applications, assuring the business continuity and uptime, and other related activities.

Managed services are not everyone's scoop. It takes a full business model in place and an extensively staffed team and well-resourced infrastructure in place to offer such services. Although they can command significant premium, they are complex to offer and manage. Generally, only seasoned and large-scale companies can afford to offer the managed services that can complement and offset such services with other types of services and offerings.

Education and Training Services

In addition to the reactive and proactive services we have discussed so far, most technology companies may also offer the educational and training services for its customers and partners. With such services, a vendor company trains its customer, directly or through its partners, on its products, technologies, and services. Additionally, the company may offer different levels of *certifications* to rank and endorse different levels of skills, after testing the knowledge and skill of the person being certified according to defined criteria. There are several types of educational and training services. For example, offering online and on-site training courses, self-learning or instructor-led, written and lab tests, and different levels of certification earnings are few common ones. In addition, training and certification can be offered along certain *tracks* of skill development to suit different market segments and customer needs.

There are several strategic and tactical advantages of offering such services. First of all, it is human nature not to like change, especially the change that is hard. Investing in training customers on a company's products and technologies creates a safety net for the business, creates a barrier to entry for the competition, and helps in customer retention over time because those customers do not generally like to retrain on newer products and new ways of doing things with another product just for the fun of it. Second, as more and more people are trained on a product and associated technologies, the knowledge base and the trained experts start defining a community in its own. This helps greatly in building and shaping a company's unique brand. Soon, the company, its products, and customer experiences are recognized through the level of trained and certified professionals it produces. Last but not the least, by offering such services, an important portion of the knowledge base and best practices are documented and preserved for repeated use. This helps shorten the learning curve for the customers, partners, and company's own employees.

Service Offering through Partners

As we had discussed in detail under the go-to-market strategy in earlier chapters, there are many types of channel partners. Since services are a high-margin business, quite often the channel partners may want to define and offer their own services wrapped around the vendor's product. In some cases, the partner services are augmented with the vendor services. Based on this, there are three broad services offering models through the channel partners, a *pass-through* model, a *detached* model, and a *combined* model. With the pass-through model, a partner only resells products and passes the service offering responsibility to the product vendor company. This allows the partner to focus on its core business while maintaining the direct customer relationships.

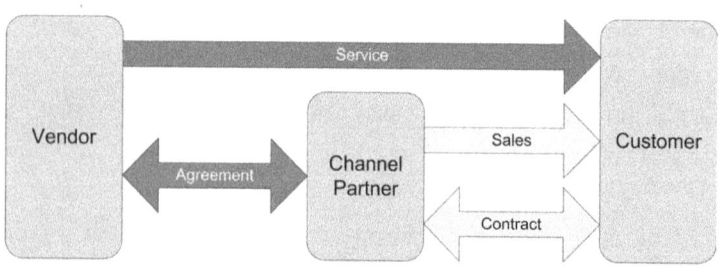

Diagram 4: Pass-Through Service Offering Model

With the detached model, a partner offers its own brand of services and does not rely on the vendor company to provide any services. The partner maintains the direct customer relationships.

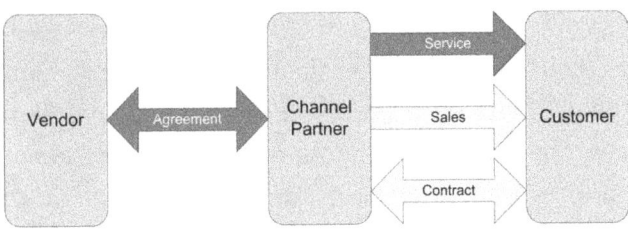

Diagram 5: Detached Service Offering Model

With the combined model, a combination of the above is used where the partner offers some services and relies on the vendor company for other services. In this way, the partner augments its services offering with those of the vendor's.

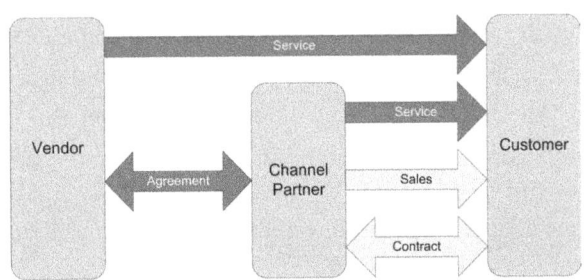

Diagram 6: Combined Service Offering Model

In case of the reseller channel partners, generally the service and support is a pass-through to the original vendor as if it was selling directly, but that may not always be the case. This may include the level-1 (initial), level-2 (advanced level), and level-3 (expert level) technical services. The service and support terms and how the end customer issues will be handled through the resell partner are agreed upon in the contract. A value-added reseller typically takes the products and provides services such as basic rack and stack, configuration and testing, and then ships the solution to the end customer for deployment. The value-added comes from the on-top or premium or professional services.

The professional services can include services such as presales network design, consulting, and post-sales training.

The system integrators maintain highly skilled talent that can provide customized solutions and tools, both hardware and software, augmented with the advanced services. Those solutions can be highly customized and built on demand. Therefore, the system integrators can charge premium for those services. The system integrators make significant revenue from the consulting services, while the rest come from the design, integration, and commissioning. The system integrators are very popular for their services in large and complex deployments where stacks are high.

In case of the OEM partners, offering high-end and high-margin branded services is one of the attractions for the OEM relationship. If an OEM partner has strong value-added services offering model, they generally want to offer their own brand of services bundled with the OEM-in products they take. The services is a high-margin business, and therefore, the OEM resellers want to leverage their own services infrastructure in place, and they can do so much comfortably when the product appears to be single vendor based to the end customer. The OEM partners also have control over the pricing for the products and services they offer and how they offer them.

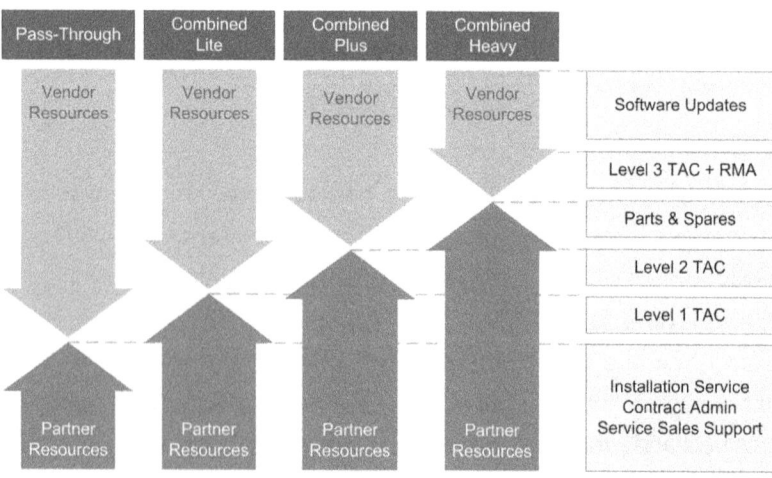

Diagram 7: Partner Service Offering Example

The diagram above shows the example of different choices and combinations of service offering through the channel partners. The partners are generally ranked into multiple tiers per a defined hierarchy to qualify for what types of service and support leverage and other incentives they can be entitled to by the vendor company. For example, the partners could be ranked as *bronze*, *silver*, or *gold* partners with gold having the privileges to most incentives. Usually, a point-scoring mechanism is provided to partners to earn reward points based on different criteria and motivate them progress up through the hierarchy. Discussing the channel incentives and reward system in detail is beyond the scope of this chapter and the book, but it gives the reader some idea.

Warranties, Returns, and Replacements

An important component of services offering is the warranties that we have not discussed until now. Whereas a service may or may not be attached to a product, a warranty is always based on a product. A warranty is a promise or assurance to receive a service such as the replacement of a product that the customer is entitled to receive should something go wrong with the product that makes it unusable. In case a product or one of its components go bad, the customer is entitled to return and replace it with an equal or better one under the warranty referred to as the *return merchandise authorizations* (RMA) in the technology world. Just like the service contracts, the warranties are valid for up to certain time; but unlike the service contracts, they usually cannot be renewed but can be extended initially.

Some companies, depending on the type of products, may choose to offer the *limited lifetime warranty* (LLW), which may or may not expire and could last for the life of the product or for a certain period after the end of life of the product—meaning, they are extended over a very long period of time. The claims under those warranties could be limited in scope though. The limited lifetime warranty could be a powerful and hard to compete with business advantage. The warranties must be fulfilled upon, or it can result in a *breach of warranty* and the legal consequences for the vendor company; therefore, a careful cost-benefit analysis should be done by the services product manager before offering such warranties.

> *A warranty is a promise to a service such as replacement of the product that the customer is entitled to receive should* something go wrong with the product that makes it unusable.

Many companies have a centralized quality organization to track and monitor the products and services quality and the customer experience. The organization uses some *key performance indicators* (KPI) to measure the quality. Statistics for the number of defects found, mean time to resolution (MTTR), number of return merchandize authorization (RMA), and the product recalls are used. The return merchandize authorization applies to collecting the malfunctioned product from the customer and shipping them a functional, brand-new replacement product. The large numbers of returns indicate the quality issues that could be rooted into the hardware or even the component level as well as the supply chain. For this reason, the technical support group maintains an inventory of spare products to be used for the returns. This obviously costs the company.

The long lead times associated with logistics and customs do not allow for the spares to be shipped worldwide from a central location while a customer's business is suffering from the downtime resulting from a product or a component failure. For the faster and global 24x7x365 service and support we discussed earlier, the vendor company usually needs to maintain the *service depots* around the world or have channel partners maintain some of those for storing the *spare parts* of the products to fulfill the return requests. For this reason, the service parts depots need to be maintained per the geographic theaters and important countries where bulk of the business is done. The overall intention is to improve the quality of service and the customer experience when they need it the most.

Key Takeaways

Depending on the company, although the product sales may drive majority of the top-line revenue for the company, the importance of services and revenue generated by them cannot be undermined. The services help improve the company's bottom-line, and for this reason, some companies may choose to move to be more service oriented rather than product centric. The shift is not easy however. As the customers are becoming more quality conscious and demand better experience, and as the demands put by the businesses become more complex, the need for advance, value-added services is rapidly increasing. This opens up great opportunities for the companies who want to capitalize on this. The services product manager can design and offer services that can really make the difference for the company's financials and for its customers.

Remember:

- A service is an intangible entity that is offered by someone and consumed by others at a given point in time, usually based on leveraging someone's expertise and usually paid for.
- The services product management means defining, creating, offering, and managing services for monetary purposes.
- The service attach rate tells someone how much of a service is being purchased as a ratio of the associated product. It is an indication of the success of the service.
- The service entitlement is the right to receive a service as a result of a contract and a service level agreement.
- The services can be divided into basic and value-added services and into presales and post-sales services.
- The professional services are usually based on hard-to-find expertise offered by highly skilled professionals and provide highly tailored services. They can be a high-margin business.
- A warranty is a promise to a service such as replacement of the product that the customer is entitled to receive should something go wrong with the product that makes it unusable.
- The return merchandize authorization (RMA) applies to collecting the malfunctioned product from the customer and shipping them a functional, brand-new replacement product.

CHAPTER-11

PROFIT AND LOSS MANAGEMENT

Financial Performance

Throughout this book, we have been talking about several financial terms such as *margins, discounts, costs, revenues, profitability*, etc., in the context of product management. It would be nice to understand in more detail what those financial concepts are and how they relate to products and services. In this chapter, we will review some fundamentals of financial performance that any product manager should be familiar with. Usually, the product manager or the leader managing the product management organization owns the *profit & loss* (P&L) responsibility for the products and services it manages. It is therefore a requirement for the senior product management role to know the financial fundamentals, have a degree in finance or business management, or have real-world experience in P&L management. Some of the key financial activities that the product manager deals with include pricing, discount structuring, cost accounting, margin analysis, revenue forecasting, and business case preparation, as discussed in detail in the earlier chapters in the book. In this chapter, we will explore more about how above contributes to the product and company's P&L and why it matters.

Income Statement

There are many indicators to determine the health of a company's business. From a product manager's perspective, one such indicator that is important to track is the company's *income statement,* also called the *financial statement, *the *annual statement, or* the *profit & loss* (P&L) *statement.* The income statement tells the overall story of a business's health in terms of revenues, costs, and net profits including margins. The product manager directly contributes to the income statement through a product's generated net revenue as well as the profitability generated through the product margins.

> An income statement states a business's total revenues, expenses, and generated income.

Usually, every vendor company has a link on its website called About the Company under which there is usually a link such as the Investors Relations that holds most of the financial information about the company including quarterly earnings transcripts and income statement. Usually, there is also a Financial Information link that leads to the archives of company's quarterly results and annual income statement. It is legally required for any public company to make those financial data available for its investors and shareholders. Let us first examine and understand fundamentals of an *income statement*. Moreover, every public company files a Form 10K with the *Securities and Exchange Commission* (SEC) for reporting its earnings. Every fiscal quarter, a company reports its financial results or *earnings*. Below is an example quarterly earnings for the hypothetical technology company we have been using as example throughout this book.

EXAMPLE 1

Nubes Networks—Q4 FY13 Earnings

San Jose, CA (Marketwire)—8/5/13—Nubes Networks today announced its fourth quarter and full fiscal year 2013 that ended on July 31, 2013. Nubes achieved record fourth quarter revenues of $578 million, representing an increase of 5% year-over-year and 4% quarter-over-quarter. Revenue for fiscal year 2013 was $2,238 million, a record for the company, up 4% year-over-year. The resulting GAAP diluted earnings per share (EPS) was $0.11 for Q4 and $0.41 for fiscal 2013, on record annual net income of $195 million. Non-GAAP diluted EPS was $0.17 for Q4, the fifth consecutive quarter of year-over-year EPS growth, and $0.66 for the year.

	Q4 2013	Q3 2013
Revenue	$ 578M	$ 555M
GAAP net income (loss)	$ 54M	$ 43M
Non-GAAP net income	$ 78M	$ 67M
GAAP net income (loss) per share—diluted	$ 0.11	$ 0.09
Non-GAAP EPS—diluted	$ 0.17	$ 0.14
GAAP gross margin	62.4%	61.3%
Non-GAAP gross margin	64.8%	63.7%

GAAP operating income	$ 86M	$ 70M
Non-GAAP operating income	$ 130M	$ 108M
GAAP operating margin	14.9%	12.6%
Non-GAAP operating margin	22.5%	19.5%
Adjusted EBITDA (1)	$ 153M	$ 131M
Cash provided by operations	$ 210M	$ 113M

First of all, let us understand one term that is used a lot in the financial statements. It is called GAAP. The term GAAP stands for *Generally Accepted Accounting Principles*, a set of accounting standards and framework applied to how the financial statements are prepared, recorded, and reported. The GAAP is used by accounting firms and corporate accountants for reporting the financial earnings. The GAAP standards and rules are complex and have evolved over long period of time. The technology companies generally report earnings under both GAAP and non-GAAP accounting guidelines.

The GAAP (Generally Accepted Accounting Principles) is a set of accounting standards and framework applied to how the financial statements are prepared, recorded, and reported.

Toward the end of the fiscal year (that is four quarters), the sum of quarterly earnings goes into the income statement. An income statement, no matter how complicated it may look like, basically consists of three key sections:

Revenue
The revenue is the overall money made through the sales of products and services before any expenses are considered. A healthy business needs to grow the revenues every fiscal year, if not every quarter. This retains investors', customers' and partners' confidence in the company. If the revenues do not grow, then the business will lead to bankruptcy and die out eventually.

Expenses

Managing expenses in any business is the key to success. In a healthy business, the revenues need to grow faster than the expenses, and management needs to keep the expense under control. If reverse happens, that is, the expenses grow faster than the revenues, and then a *runaway budget deficit* happens. The expenses include all costs of doing business including *research and development* (R&D) costs, employee salaries, administrative costs, facilities, supplies, costs of marketing and sales, etc.

Net Result

The net result that a business has produced shows how well the business has done during the given period. The net result can be obtained by subtracting the total expenses from the total revenues. A *net income* has occurred if the revenues were more than the expenses, or a *net loss* has occurred otherwise.

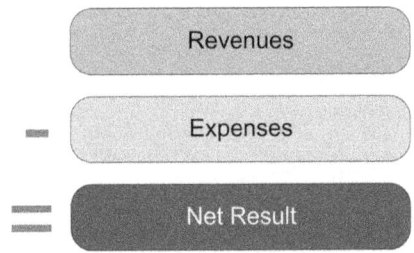

Diagram 1: Basic Structure of an Income Statement

Below, we will analyze the income statement for the hypothetical technology company we have been using as example throughout this book. Let us explore some of the key information in this financial statement next.

EXAMPLE 2

Nubes Networks – 2013 Consolidated Income Statement

Revenues	Net Revenues	$25,150,995
	Cost of Revenue	$9,570,455

	Gross Margin	$ 15,580,540
Operating Expenses	Research & Development	$4,250,585
	Sales & Marketing	$2,508,052
	General & Administrative	$650,379
	Amortization of Intangible Assets	$539,351

	Total Operating Expenses	$7,948,367
Income	Income from Operations	$4,170,520
	Interest Income	$50,250
	Other Expenses	($95,455)

	Net Income (Loss)	$4,125,315
Earning	Net Income (Loss) per Share (Basic)	$0.70
	Net Income (Loss) per Share (Diluted)	$0.68
	Shares used (Basic)	5,875,050
	Shares used (Diluted)	5,995,343

Overall P&L Summary Section

At the very top in an income statement are the *net revenues, cost of revenues,* and *gross margin* information as the overall summary for anyone interested to know without reading the details below. This is the overall P&L summary.

- **Net Revenue** is the revenue that the company actually made after the products or services discounts before subtracting the expenses yet. As we recall from the product pricing discussed earlier, the actual price at which the product or service generally sells is the list price minus discount, and that is the average sale price (ASP). The net revenue, similarly, is the revenue made through all products or services after the applicable discounts. The revenue made through all the hardware and software product sales, licenses, royalties, and the revenue made through all the services sales such as basic, premium, professional, and educational services, everything is included under this entry.

- **Cost of Revenues** is the total cost that was incurred by the company to build the above products and to create the above services to generate the above revenue. This includes, for example, the total of all costs of goods sold (COGS) for the products. This is the materials cost that was spent on building or acquiring those products. The cost of revenues also includes product spares and return merchandize authorization (RMA) related costs, logistics, and other such costs used to offer the services.

- **Gross Margin** is the difference between the net revenues and the cost of revenue. In other words, gross margin indicates the profitability that the products and the services generated before other expenses are yet considered.

Gross Margins = Net Revenue – Cost of Revenue

Detailed Operating Expenses

The operating expenses indicate all expenses that the company bore in bringing the above products and services to the market and generating the revenue reported earlier. This includes employee salaries, money spent on research and development, filing patents, marketing campaigns and programs, sales activities and travels, administrative costs, and more. Keeping the operating expenses low is always a challenge that the company executives have to deal with. The lower operating costs indicate how productive and efficient a company is in terms of running its operations.

Below are some of the common expense types:

- **Research & Development (R&D) Costs** are the costs that the company spends on technological research and developing in building the new hardware or software products including the salaries for engineers, engineering capital costs, prototype costs, patent fees, testing and qualification costs, and more. This includes the nonrecurring engineering (NRE) costs we discussed during the product planning phase. The innovative and growing companies set aside decent budget for the R&D activities, and this is one of the indicators for a promising company future. Great products are the most important thing for a hi-technology product—period. When it comes to the cost cutting, this should be the last area to cut the budgets. However, there is always the room to optimize the costs and productivity.

- **Sales and Marketing Costs** result from the money spent on the sales and marketing activities such as the marketing development fund (MDF), advertising expenses, customer events, trade shows, printed material, partner promotion activities, trade shows and events spending, customer meetings and entertainment, travel, sales cash incentives, other such costs, etc. Keeping marketing and especially sales costs under control is very important for a company because those costs can quickly spike from quarter to quarter and are hardest to track. If sales effort does not result into proportional bookings, then it is purely the money wasted.

- **General and Administrative Costs** result from everyday administrative activities such as printing, office mail, office supplies, break room supplies, and many other such costs that add up quickly. The administrative costs are usually not the biggest costs, but they are important to control.

- **Legal Costs** are the costs involving money spent on legal fees and court proceedings dealing with lawsuits or litigations, Securities and

Exchange Commission (SEC) filings, etc. We discussed the importance of legal disclaimers and liabilities under the product roadmap earlier that could result in legal costs to the company.

- **Other Costs** could list several other types of costs not covered under any of the above categories.

- **Total Operating Expenses** is the sum of all of the above expenses that the company has spent on running the business. As discussed earlier, if the operating expense grows at higher rate than the revenues, then the income will shrink and the company will head toward a financial disaster.

Detailed Income Section

The income section of the statement lists all types of income that the company generated from different sources over the reported period. Basically, this is the *savings* or cash that business has generated. If the business is not generating enough income every year, then there is not much justification for the business.

- **Income from Operations** or also called the *operating income* is the cash that the company has generated before paying any *interest and taxes* (EBIT). The inventors track the operating income as an indicator of a business's earning potential. Operating income is derived by subtracting the operating expenses and the cost of revenues from the total revenues:

    Operating Income = Net Revenue – (Cost of Revenue + Operating Expenses)
    Operating Income = Gross Margins – Operating Expenses

- **Interest and Other Income** is the income generated from other sources other than selling the products and services, such as any investments company had made, interest, or rental income generated from assets, etc. Basically, this could include the income that is generated from other than the company's core business in this case.

- **Net Income (or Loss)** is the most important part of the income statement, and hence the statement is named after it. The net income is the sum of income from all sources, if it is a positive number; otherwise, it is the net loss. Whereas the net revenue indicates the "top line" of the business, the net income indicates the "bottom line" of the business and indicates the net profit business has made. The company may pay *dividends* from the net income and carry over the remaining to the *retained income* section on its *balance sheet*.

Earnings Section

The earnings section explains what each shareholder of the company, if it is a public company, has made or lost. One of the most important indicators of a business's earning potential is the *earning per share* (EPS), which is derived by taking the net income and distributing it among all shares issued:

Earnings per Share = Net Income / Number of Outstanding Shares

Return on Sales

There is another measure of the financial performance that is commonly used. It is called the *return on sales* (ROS). It is an indication of the efficient business operations providing a picture on how efficiently a company controls its operating costs to boost the profitability. Therefore, it is also referred to as the *operating margin*. Companies with high return on sales can handle tough economic times and downturns much better. The higher the return on sales, the better the business for investors and shareholders as it tells how much of each dollar of sales can be turned into the profits. The return on sales is calculated as the percentage of net income (profit) compared to the net revenue (sales):

Return on Sales = Net Income (profits) / Net Revenue (sales) x 100

The return on sales can be used to compare the business efficiency of the vendor company by itself or to compare it with other companies in similar businesses. The return on sales can also be used to measure the success of a business model such as the OEM-in and resell. This makes sense since in the case of the OEM-in and reselling, there are no products built, so there are no R&D costs associated. In this case, the products are taken and sold, incurring mostly sales and marketing related (go-to-market) costs. Therefore, the return on sales provides the best measure of how successful the build-versus-buy-versus-partner decision and the OEM or resell business have been.

> The return on sales (ROS) is an indication of the efficiency of the business operations of a company.

Balance Sheet

Another important financial statement and health indicator of a business is its *balance sheet*. Whereas the income statement sums up the conclusive financial picture of a business at the end of a fiscal period, a balance sheet shows a snapshot of a business at any given point in time. The balance sheet has three basic sections: the *assets*, the *liabilities,* and the *shareholders' equity*. The income statement and the balance sheet together show a business's financial position. The balance sheet can be represented by the following equation:

$$Shareholders' \ Equity = Assets - Liabilities$$

> A balance sheet shows a snapshot of a business at any given point in time and how balanced it is including its assets, liabilities, and equity.

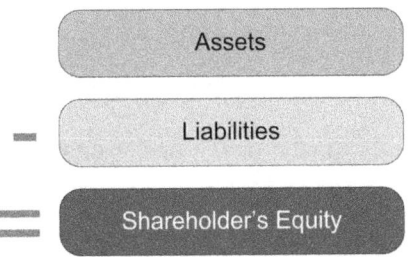

Diagram 2: Basic Structure of a Balance Sheet

Assets

The asset part of the balance sheet includes all assets that the business owns, including the liquid assets such as cash, as well as property and other assets.

Generally, this includes only the *current assets*, that is, the assets having a life of less than a year, such as cash, cash equivalent, accounts receivables, and any product inventory. The *noncurrent assets* include items such as the plant, property, equipment, any investments as well as intangible assets based on the intellectual property (such as the software and patents) and goodwill, etc. Assets are critical in keeping a business going.

Liabilities

The liabilities section of the balance sheet includes what the business owes to others. This may include the loans, bonds, and accounts payables. The higher the liabilities, the worse the business situation is.

Shareholders' Equity

This is the portion of a balance sheet that shows how the ownership and income of the company are distributed among its shareholders in the form of *shares* or *stocks*. It is not necessary that all the income be distributed. The portion of the income that is kept in the company account is referred to as the *retained income*. A *retained loss* is the opposite, that is, a loss that is retained by the company. The retained income and the shareholders' equity together indicate how much of the business can be funded for reinvesting into the areas of strategic importance without borrowing money from the investors.

EXAMPLE 3

Nubes Networks – 2012 Consolidated Balance Sheet

Assets	Cash, Cash Equivalent, Investments	$50,668,243
	Working Capital	$11,377,250
	Total Assets	$350,674,075
Liabilities	Loans, current and long term	$5,089,544
	Noncurrent liabilities	$97,250
Shareholders' Equity		$387,586,352

Product's and Services' Contribution to P&L

Now that we have analyzed in detail the two most important financial pieces of information regarding a business's financial health—the income statement and the balance sheet—we can better relate and understand how a product manager can contribute to the overall financial performance through its products and services related decisions and activities—that is, how a product and service level P&L can contribute to the overall P&L of the company.

Contribution to the Top Line

How much revenue a product or service generates directly contributes to the net revenues on the company's income statement. How the product manager sets the pricing of the product or the service, how the discount structure is determined, what the average sale price comes out to be, all add up to the net revenue of the product or the service. This adds up to the net revenue of the company on the income statement, which is the sum of all net revenues from multiple products and services sales. As discussed in the previous chapters, if the price positioning is not done right, the product and attached service revenues will be affected, which means that the company's revenues will be affected.

Other factors discussed during the product and service life cycle phases can also affect the revenues, such as a too early or too late launch of the product or the service. Therefore, pretty much every decision the product manager and executives make affects the company's top line. Below, we will revisit the important tasks from the product and service life cycle and relate them to the overall P&L impact so that it becomes clear how important it is, that a product manager does for the business.

Market Opportunity

As we discussed during the product planning phase, the very first step that the product manager takes to come up with a product idea and what target markets it will be sold into will later on affect how much revenue the company could make at best. Until a product lives its life, all the revenue generated by it is because of the idea that might have been originated by the product manager. Picking up a high-growth market with sizable total addressable market or picking up a modest size but highly realizable market with least competition both could turn out to be good choices later on in terms of the top-line growth, provided that everything else discussed in this book is executed upon well.

For example, if the total addressable market is $10 billion and there are ten major competitors, it is still a sizable market to gain the market share. If the total addressable market is $1 billion with only one other competitor, that is not bad either. However, if the total addressable market is $1 billion with ten major competitors, perhaps it may not be a good idea to enter that market, and the next level of details will need to be evaluated. Therefore, choosing the right market, when coming up with the new product idea, does impact the company's top line later on. In case of service offering, the situation is similar. Picking up the right market, getting the right set of customers, and defining the right services will greatly affect the service revenues.

Product Positioning

Creating the right product positioning is one of the most important activities for the fate of a product. Wrong product positioning can cause failure in the marketplace, even if the product is first in time to market. The product positioning draws the circle around the product within which the product is intended to be used. If the product manager wisely positions the product at the right market tier and for the right use cases, the product will likely be successful in generating the revenues and the profits desired. The product positioning also dictates what type of pricing and margins the product will command depending on the role it will be sold in. For example, in case of networking, it may be harder to command high average sale prices in case of a Top-of-Rack switch compared to a core switch. Or it may be easier to command higher high average sale prices on the same switch product in the data center versus enterprise campus market. Therefore, the product positioning can also affect the company's top line.

Time to Market

The product manager should have a clear idea when a product or service should get ready and launch as this may greatly affects the business outcome.

A late product, even the right one, may lose its effectiveness in terms of surprise factor, market lead, and the competitive advantage if the competition announces a product sooner. Similarly, being first to market is also not always good, but there are situations when it is important. In case of service offering, timing also does matter, especially when there are other vendors who could offer similar or better services. Releasing a product at the right time in the market has the opportunity to grab the market share before the competition can catch up. This helps the company's top-line revenue.

Product and Service Pricing

Pricing has an extremely important role in the success of a product or service since in almost every sales opportunity it is one of the most important, if not the topmost, considerations for the end customer's purchasing decision. Pricing the product just right pays off for long time and affects both top line and bottom line of the company. If a product is priced too high, it will result in customers not buying it at all or not buying it much. This will result in less net revenues. On the other hand, if a product is priced too low, it may sell in decent volumes but will not generate the net revenues and the profits it could have otherwise per its potential. So the revenue is "truncated."

The services are affected in two ways. First, if the product pricing is not right, the low volume of the product will also pull down the service attach rates. Second, if the service pricing itself is too high, it will discourage customers to buy those services. Additionally, as we discussed in detail in earlier chapters, the product manager also needs to price the products and services accordingly to compensate for the higher regional discounts and still sell at the target average sale price and margins. The pricing should be tuned regionally such that the product or the service sells at the highest possible average sale price, since it is the average sale price that directly equates to the net revenue on the income statement.

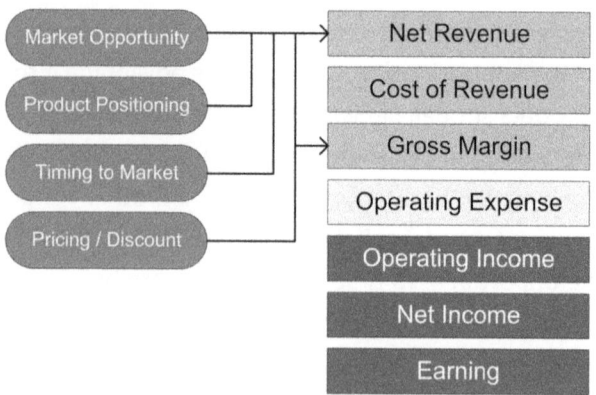

Market Opportunity		Net Revenue
Product Positioning		Cost of Revenue
Timing to Market		Gross Margin
		Operating Expense
Pricing / Discount		Operating Income
		Net Income
		Earning

Diagram 3: Product Contribution to Overall P&L Top-Line

Contribution to the Bottom Line

The way a product is designed and built, its material, manufacturing, and other costs as well as the discount structure will determine the product margin. The costs mentioned above directly add up to the cost of revenue, while the product margins directly contribute to the gross margins on the income statement. The gross margins directly influence the operating and net incomes. The program costs to develop the product and the recurring costs to sustain it add up to the operating expenses and reduce the operating income.

For the services, the costs such as the cost of providing technical support and replacement parts add up and influence the service margins and the gross margins overall. The other costs such as the services team salaries impact the operating expenses. Services are generally higher margin and lower cost business compared to the products and can greatly improve a company's bottom line. The product manager, operations, manufacturing, engineering, and services teams all can help improve the bottom line of the business since they can influence the factors outlined above to certain degree.

Product and Service Costs

The product margins impact the profitability directly. If we remember for the earlier chapters, the product COGS depend on several factors. First, the raw material cost and the costs negotiated with the suppliers based on the demand forecast impacts the gross margins since it translates into the cost of revenue. Second, keeping the manufacturing overhead low also helps in keeping the COGS low. Therefore, keeping the burdened-COGS low is very important. By the same token, if the product quality is good and the *mean time between failures* (MTBF) on its component is longer, the cost of return merchandize can be lowered resulting in better service margins. If the cost of providing technical support such as time spent on per-support call is less, it can also help in lower service offering cost (operating costs) and better margins. The lower the product or service costs, the lower the cost of revenue on the income statement and the higher the gross margin and the operating income.

Product and Service Pricing
As discussed earlier, if a product is priced too low, it may sell in decent volumes but with low margins since there are still discounts involved. If the product is priced too high, it will not sell in desired volumes. Eventually, its demand will start declining, and it will become very difficult for the commodity team to maintain the required level of COGS with the suppliers, further eroding the margins. And if the product volume goes down, so do the attached services and the costs of offering those services go up, reducing the service margin as well. Either situation above affects the product and service margins, and therefore the gross margins and the operating and net incomes on the income statement. As discussed earlier, pricing the product or service at the right level on regional basis such that it sells at the best possible average sale price promises more income for the business.

Program Costs
As we had discussed earlier, a significant program cost component involved in building a product is the nonrecurring engineering cost. Those costs include salaries spent on the resources working on the project, capital equipment costs, compliance costs, and other such items. Not only the type and design of the product may affect the program costs but also what type of resources is hired to build the product and where the work is performed. The burden of program cost control primarily lies on the core team and the project manager. The program costs directly contribute to the operating expenses on the income statement and therefore reduce the operating income by that factor.

Maintaining highly productive teams, good resource management leadership, and distributing the work are few of the factors that can help getting more done with less.

Sales and Marketing Costs
One of the major sources of operating costs comes from the money spent on sales and marketing activities such as marketing development fund, advertising expenses, customer events, trade shows, printed material, partner promotion activities, customer meetings and entertainment, customer tours, sales and partner incentives, travels, and other such costs. Keeping marketing and especially sales costs under control is very important for a company because sales costs can quickly spike from quarter to quarter and are hardest to track. If sales effort does not result into proportionally rising product bookings, then it is purely the money wasted. Similarly, the marketing spend has to be tackled thoughtfully. Instead of being present everywhere one can, having focused and targeted marketing events may be a wiser strategy under tighter financial situations. Having fewer but high-quality and focused content may be more effective than lots of mediocre content that no one reads and so forth. Such costs can be greatly controlled by building and maintaining highly efficient and productive teams that consists of top talent. Having fewer but well-rewarded, top-skilled resources can actually save costs to the company in longer run than having lots of unmotivated, mediocre resources and laying them off frequently.

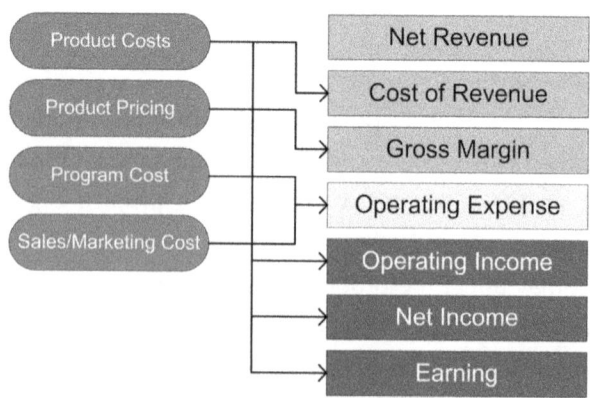

Diagram 4: Product Contribution to Overall P&L Bottom-Line

Whereas the product manager and the core team can directly influence the income statement of the company as explained above, they have little to do with the company's balance sheet since they do not decide which assets to acquire, where to invest, or whether to take a loan or not. Those types of decisions reside with the executive team and the board of directors.

Key Takeaways

The product manager's contribution to the company's financial performance is undisputed, and if someone is at unrest all the time trying to make the business successful, it is the product manager. The product manager's challenges are unending, its responsibilities are rigorous, and what it does has direct impact on turning a company from good to great. If there are no products, there are no sales, and hence there is no revenue. Neither there is any profit to make. Still, this is a role least in the spotlight or much appreciated. It would be fair to state that for every product unit sold, the product manager should be compensated for; however, this almost never happens. The product manager is a self-driven, self-motivated, highly passionate personality that goes on fueled by the spirit of entrepreneurship, advancement, and winning.

Remember:

- An income statement states a business's total revenues, expenses, and generated income.
- The balance sheet shows a snapshot of a business at any given point in time and how balanced it is including its assets, liabilities, and equity.
- The return on sales is an indication of the efficiency of the business operations of a company.
- In a healthy business, revenues need to grow faster than the expenses.
- If sales effort does not result into proportionally growing product bookings, then it is purely money wasted.
- A product or service P&L contributes to the overall P&L of the company directly.

- How much revenue a product or service generates directly contributes to the net revenues on the company's income statement and impact the top line.
- The way a product is designed and built, its material, manufacturing, and other costs as well as the discount structure determine the product margin and impact the bottom line.
- Services are generally higher margin and lower cost business compared to the products and can greatly improve a company's bottom line.

Summary

Product management is a demanding but exciting career. The product manager's challenges are unending, his responsibilities are rigorous, and what he does, has direct impact on a company's financial performance. Building and launching new products and turning an idea from a piece of paper into a functional product is almost a miracle. In addition, the product manager manages the product throughout its life. In doing so, the product manager deals with pretty much every function in the company.

Speaking of the product life, anything done well during the planning phase will pay off during the other phases of the product life cycle. The execution phase is the phase when a product really takes shape. Once the product is complete and ready to be launched, it is an exciting time for the product manager. The product is ready to put under real-world test.

Just building and launching a product is not enough. Target customers should be told about how great a product is, which takes good marketing and evangelism. Market routes must be established to sell and promote the product and make business out of it. Additionally, different types of services can be defined to be attached with the product as an overall offering. Defining and implementing a go-to-market plan for the product is complicated but interesting set of activities. If the go-to-market ecosystem is set up well, the product manager can watch his product's and associated services' revenues multiply.

Once the product is out there, it needs to be taken care of. Sustaining a product takes effort. This is the time to turn a good product into a great product to take the product toward completeness and maturity. Eventually, any product will get old and obsolete. Even the greatest of products must be given a farewell, and the end of life must happen to keep the innovation wheel rotating. New products and services enter the picture, and the product management action starts all over again.

Index

A

accounting books, 145, 147, 183, 190
account management, 172
advanced services, 225, 245
agile methodology, 41, 83
AM (account managers), 142, 258
analysis
  build-versus-buy-versus-partner, 62,
    222, 227, 230
  competitive, 64, 84, 142, 259-60, 266
analyst relations and public relations,
  135, 232
animations, 250
API (application programmable
  interfaces), 97
approach, hit-and-trial sales, 260
Architecture Design, 93
ASP (average sale price), 57, 59, 119-20,
  124, 270, 291, 297, 299
attach rates, 268
awareness, 132-33, 146, 164, 231, 234,
  254

B

balance sheet, 32, 160, 294-97, 303
BD (business development), 24, 35, 172,
  226, 254, 261
BDM (business development manager),
  142, 258
benchmark, 57, 117, 130, 140, 152, 156,
  239, 242, 249
Benchmark Testing, 118
Benchmark Test Report, 139-40, 249
beta customers, 101-2
beta trial, 69, 99-102, 117-18
board design, 95, 100
BoM (bill of material), 56, 109, 112
bookings, 106, 128-29, 145, 160, 168-69,
  173, 187, 202-3, 216

bottom line, 193, 205, 269, 278, 294, 299-
  300, 304
bottom-up, 122, 124
breach of warranty, 283
Breakeven, 84, 86
breakeven point, 85-86, 114
briefings, 133, 137, 232, 248, 252-53
bug fixing, 101
burdened-COGS, 56, 122, 301
business case, 22, 58-59, 62, 68, 74, 84-
  87, 114, 152, 197, 228, 286
business development teams, 134, 170,
  172, 234

C

CAM (channel account managers), 142,
  258
CC (concept commit), 42, 58-59, 79, 84, 87
CEO (chief executive officer), 20, 38, 85,
  240
certification, 72, 279
CFD (customer found defects), 156, 158
channel conflict, 208, 214, 218, 224-25,
  230
channel partners, 108-9, 120, 127, 138,
  164-66, 170-71, 187, 207-8, 210-12,
  214-19, 221-22, 224-27, 230, 242,
  258, 261, 278, 280, 283-84
channels, 19, 30, 39, 165, 204, 206, 208,
  212, 214, 217, 225-26, 229-30
channel sales, 19, 170, 205-7, 216-18,
  220, 229, 261
CLI (command line interface), 91, 143
CMO (chief marketing officer), 35, 193
COGS (cost of goods sold), 56-59, 70, 85,
  122, 124, 188, 213, 269, 291, 300-301
cold-calling, 257
collateral damage, 99, 104, 274
combined model, 225
commodities, 109, 188

Product Management

industry events, 137, 140-41, 250-51, 254

initial BoM request, 111

intent, plan of, 149-50, 193-99, 203

interest, conflict of, 206, 217

Interest and Other Income, 293

inter-product positioning, 51

intra-product positioning, 51

inventory, 107, 160-63, 180-81, 183-85, 187, 189, 220, 262, 284

inventory management, 107, 112, 115, 161-62, 179, 215, 221

Inventory Snapshot, 184

Investors Relations, 287

ISO (International Standards Organization), 72, 155

K

key message, 129-30, 146, 238

KPI (key performance indicators), 154, 284

L

LA (limited available), 143

landing webpage, 139

Lead generation, 234

lead time, 145, 160-62, 173, 179, 209, 215, 221, 284

Legal Costs, 292

legal disclaimers, 195, 199-201, 203, 236, 293

levels, severity, 157-58

liabilities, 203, 293, 295-96, 303

license vouchers, 57

LLW (limited lifetime warranty), 283

LMF (layered messaging framework), 130, 240

lost opportunity cost, 74, 78, 198

LP (list price), 47, 57, 119, 122, 267, 269-70

LPM (launch project manager), 134

M

maintenance release, 157, 182

major release, 157, 182

managed services, 278-79

manufacturing overheads, 56-57, 59, 188, 269, 301

manufacturing part number, 109-10

manufacturing process, 20, 113-14, 155, 184, 190

marcom (marketing communications), 232-33

margin analysis, 56, 58, 84, 124, 286

margins, 56-57, 59, 85, 118-19, 122, 124, 146, 165-66, 185, 208, 218-20, 226, 267, 269-70, 286, 298-99, 301

profit, 85, 269

market, 18, 24-25, 32, 37, 40, 42-43, 45, 48-49, 51, 57, 61-65, 68-69, 74, 83, 90, 94, 112, 125, 129, 132, 141, 151, 161, 163, 168, 170, 174, 189, 191, 197, 205-8, 216-17, 221-22, 226-29, 248-51, 253, 261, 291, 298-99

realizable, 48, 53, 59, 78, 84, 298

market acceptance, 42

marketing, 7, 13, 21, 24, 30, 37, 66, 108, 110, 116, 132, 135, 140, 142, 163-65, 173, 191, 193, 224, 226, 231-32, 234-36, 250, 258, 262-63, 265-66, 289, 292, 295, 302

marketing collateral, 131, 134-35, 236, 238-39

marketing events, 220, 226, 261, 302

marketing forecast, 70, 106, 115, 161

marketing manager, 134, 137, 166, 231, 237-38, 242, 252, 254, 260, 262

marketing part numbers, 68, 109

marketing program or project manager, 237

marketing strategy, 122, 141, 204, 229, 251

marketing teams, 34, 131, 137-38, 164, 166, 233-34, 236-37, 244, 251, 253-54, 256, 260-61

market opportunity, 42, 47, 59, 64, 142, 258, 297

final, 48

Market Opportunity Overview, 64

market requirements, higher-level, 78

top-down, 122

top line, 193, 294, 297-99, 304

total cost, 56-57, 85-86, 114, 136, 266, 291

Total Operating Expenses, 290, 293

TQM (total quality management), 80

traceability, 71-72

trade show, 25, 135, 137, 232, 254-55, 292, 302

training services, 279

trouble tickets, 102, 273

truncated, 299

TSS (technical support service), 268, 271-72, 274

TTM (time to market), 25, 69, 83, 94, 298

Two-Tier Channel Fulfillment Model, 218

V

value-added services, 206, 210, 225, 235, 269, 271, 285

value proposition, 51, 118, 131, 135, 137, 140, 142, 146, 205, 236, 239-40, 243, 248-49, 253, 258

VAR (value-added resellers), 210-12, 215, 218, 220, 225, 230, 281

vendor company, 113, 155, 174, 182, 187, 200, 207, 209, 212, 215, 224-25, 227, 250, 272, 281, 283-84, 287, 294

video data sheet, 140, 249-50

W

warranties, 128, 270, 283-85

waterfall method, 41, 80-83

CPSIA information can be obtained
at www.ICGtesting.com
Printed in the USA
BVHW030918070619
550448BV00005B/92/P